Watershed and Drainage

Watershed and Drainage

Drick Deep

Oxford Book Company
Jaipur ▪ New Delhi

Published By

Oxford Book Company

Regd. Off. "Bony Residency", Gate No. 2
Opp. Tilak Public School, Vishweswariya Nagar,
Gopalpura Road, Jaipur - 302018, India
Ph: 0141 - 2761280, Telefax: 2761381
email - oxfordbookjpr@gmail.com
website - oxfordbookcompany.com

Delhi Office: 102, 1st Floor, Satyam House,
4327/3, Ansari Road, Darya Ganj,
New Delhi - 110002, India
Ph: 011-45652440

Published in India, 2025

ISBN : 9789380179421

Price : ₹ 2175/-

Typeset by Shivangi Computers, Jaipur

Printed and bound in India by Thomson Press Ltd.,
New Delhi

Preface

The present book is a practical guide to all the personnel working on watersheds, at different levels. Though the main thrust in the book has been soil and water conservation; which are very important aspects in watershed management; the socio-economic factors and many other aspects have been considered.

The contents not only encompass, whether it is related to watershed characterisation, watershed requirements, land development, livestock development, organic farming systems, silvipastoral approach, renewable energy sources or homestead farming systems but also development of people in the watershed. The whole gamut of watershed is available in this book.

Everybody engaged in watershed related study or working in the watershed areas could find answers for all his/her problems. Species of vegetation suitable for plantations, whether it is in agricultural lands or non-arable lands or in drainage line treatment, are specifically indicated in this book. Their adaptability in several agro-climatic conditions also has been specified. Design, measurements and modes of execution of soil and water conservation structures have also been elaborately mentioned in the book."

Drick Deep

Contents

Chapter 1

Introduction

WATER RESOURCES

Water resources are sources of water that are useful or potentially useful to humans. Uses of water include agricultural, industrial, household, recreational and environmental activities. Virtually all of these human uses require fresh water. 97.5% of water on the Earth is salt water, leaving only 2.5% as fresh water of which over two thirds is frozen in glaciers and polar ice caps.

The remaining unfrozen freshwater is mainly found as groundwater, with only a small fraction present above ground or in the air. Fresh water is a renewable resource, yet the world's supply of clean, fresh water is steadily decreasing. Water demand already exceeds supply in many parts of the world, and as world population continues to rise at an unprecedented rate, many more areas are expected to experience this imbalance in the near future. The framework for allocating water resources to water users (where such a framework exists) is known as water rights.

SOURCES OF FRESH WATER

SURFACE WATER

Surface water is water in a river, lake or fresh water wetland. Surface water is naturally replenished by precipitation and naturally lost through discharge to the oceans, evaporation, and sub-surface seepage. Although the only natural input to any surface water system is precipitation within its watershed, the total quantity of water in that system at any given time is also dependent on many other factors. These factors include storage capacity in lakes, wetlands and artificial reservoirs, the permeability of the soil beneath these storage bodies, the runoff characteristics of the land in the watershed, the timing of the precipitation and local evaporation rates. All of these factors also affect the proportions of water lost.

Human activities can have a large impact on these factors. Humans often increase storage capacity by constructing reservoirs and decrease it by draining wetlands. Humans often increase runoff quantities and velocities by paving areas and channelizing stream flow. The total quantity of water available at any given time is an important consideration. Some human water users have an intermittent need for water. For example, many farms require large quantities of water in the spring, and no water at all in the winter. To supply such a farm with water, a surface water system may require a large storage capacity to collect water throughout the year and release it in a short period of time.

Other users have a continuous need for water, such as a power plant that requires water for cooling. To supply such a power plant with water, a surface water system only needs enough storage capacity to fill in when average stream flow is below the power plant's need. Nevertheless, over the long term the average rate of precipitation within a watershed is the upper bound for average consumption of natural surface water from that watershed. Natural surface water can be augmented by importing surface water from another watershed through a canal or pipeline. It can also be artificially augmented from any of the other sources listed here, however in practice the quantities are negligible. Humans can also cause surface water to be "lost" (i.e. become unusable) through pollution. Brazil is the country estimated to have the largest supply of fresh water in the world, followed by Russia and Canada.

BELOW WATER

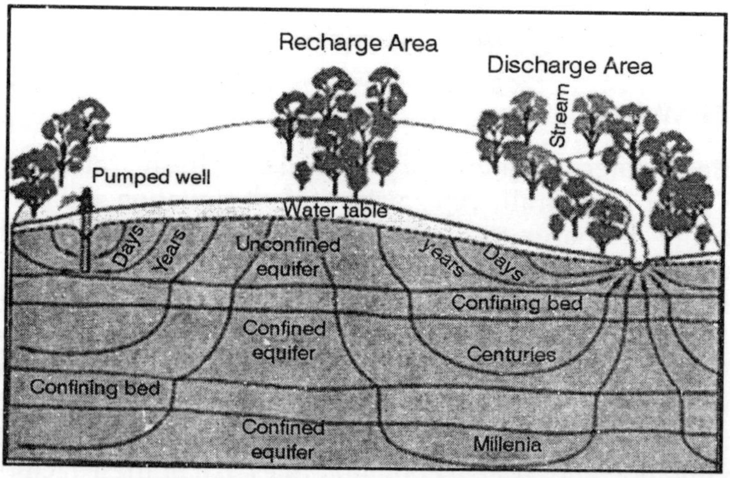

Fig. Sub-Surface Water Travel Time

Sub-surface water, or groundwater, is fresh water located in the pore space of soil and rocks. It is also water that is flowing within aquifers below the water table.

Sometimes it is useful to make a distinction between sub-surface water that is closely associated with surface water and deep sub-surface water in an aquifer (sometimes called "fossil water"). Sub-surface water can be thought of in the same terms as surface water: inputs, outputs and storage. The critical difference is that due to its slow rate of turnover, sub-surface water storage is generally much larger compared to inputs than it is for surface water. This difference makes it easy for humans to use sub-surface water unsustainably for a long time without severe consequences.

Nevertheless, over the long term the average rate of seepage above a sub-surface water source is the upper bound for average consumption of water from that source. The natural input to sub-surface water is seepage from surface water. The natural outputs from sub-surface water are springs and seepage to the oceans. If the surface water source is also subject to substantial evaporation, a sub-surface water source may become saline. This situation can occur naturally under endorheic bodies of water, or artificially under irrigated farmland. In coastal areas, human use of a sub-surface water source may cause the direction of seepage to ocean to reverse which can also cause soil salinization. Humans can also cause sub-surface water to be "lost" (i.e. become unusable) through pollution.

Humans can increase the input to a sub-surface water source by building reservoirs or detention ponds. Water in the ground is in sections called aquifers. Rain rolls down and comes into these. Normally an aquifer is near the equilibrium in its water content. The water content of an aquifer normally depends on the grain sizes. This means that the rate of extraction may be limited by poor permeability.

Desalination

Desalination is an artificial process by which saline water (generally sea water) is converted to fresh water. The most common desalination processes are distillation and reverse osmosis. Desalination is currently expensive compared to most alternative sources of water, and only a very small fraction of total human use is satisfied by desalination. It is only economically practical for high-valued uses (such as household and industrial uses) in arid areas. The most extensive use is in the Persian Gulf.

Frozen Water

Several schemes have been proposed to make use of icebergs as a water source, however to date this has only been done for novelty purposes. Glacier runoff is considered to be surface water.

USES OF FRESH WATER

Uses of fresh water can be categorized as consumptive and non-consumptive (sometimes called "renewable"). A use of water is consumptive if that water is not immediately available for another use. Losses to sub-surface seepage and evaporation are considered consumptive, as is water incorporated into a product (such as farm produce). Water that can be treated and returned as surface water, such as sewage, is generally considered non-consumptive if that water can be put to additional use.

Agricultural

It is estimated that 69% of world-wide water use is for irrigation, with 15-35% of irrigation withdrawals being unsustainable. In some areas of the world irrigation is necessary to grow any crop at all, in other areas it permits more profitable crops to be grown or enhances crop yield. Various irrigation methods involve different trade-offs between crop yield, water consumption and capital cost of equipment and structures. Irrigation methods such as most furrow and overhead sprinkler irrigation are usually less expensive but also less efficient, because much of the water evaporates or runs off. More efficient irrigation methods include drip or trickle irrigation, surge irrigation, and some types of sprinkler systems where the sprinklers are operated near ground level. These types of systems, while more expensive, can minimize runoff and evaporation. Any system that is improperly managed can be wasteful. Another trade-off that is often insufficiently considered is salinization of sub-surface water.

Aquaculture is a small but growing agricultural use of water. Freshwater commercial fisheries may also be considered as agricultural uses of water, but have generally been assigned a lower priority than irrigation. As global populations grow, and as demand for food increases in a world with a fixed water supply, there are efforts underway to learn how to produce more food with less water, through improvements in irrigation methods and technologies, agricultural water management, crop types, and water monitoring.

Industrial

It is estimated that 15% of world-wide water use is industrial.

Major industrial users include power plants, which use water for cooling or as a power source (i.e. hydroelectric plants), ore and oil refineries, which use water in chemical processes, and manufacturing plants, which use water as a solvent. The portion of industrial water usage that is consumptive varies widely, but as a whole is lower than agricultural use.

Household

It is estimated that 15% of world-wide water use is for household purposes. These include drinking water, bathing, cooking, sanitation, and gardening. Basic household water requirements have been estimated by Peter Gleick at around 50 liters per person per day, excluding water for gardens.

Recreation

Recreational water use is usually a very small but growing percentage of total water use. Recreational water use is mostly tied to reservoirs. If a reservoir is kept fuller than it would otherwise be for recreation, then the water retained could be categorized as recreational usage. Release of water from a few reservoirs is also timed to enhance whitewater boating, which also could be considered a recreational usage. Other examples are anglers, water skiers, nature enthusiasts and swimmers.

Recreational usage is usually non-consumptive. Golf courses are often targeted as using excessive amounts of water, especially in drier regions. It is, however, unclear whether recreational irrigation (which would include private gardens) has a noticeable effect on water resources. This is largely due to the unavailability of reliable data. Some governments, including the Californian Government, have labelled golf course usage as agricultural in order to deflect environmentalists' charges of wasting water. However, using the above figures as a basis, the actual statistical effect of this reassignment is close to zero.

Additionally, recreational usage may reduce the availability of water for other users at specific times and places. For example, water retained in a reservoir to allow boating in the late summer is not available to farmers during the spring planting season. Water released for whitewater rafting may not be available for hydroelectric generation during the time of peak electrical demand.

Environmental

Explicit environmental water use is also a very small but growing percentage of total water use. Environmental water usage includes artificial wetlands, artificial lakes intended to create wildlife habitat, fish ladders around dams, and water releases from reservoirs timed to help fish spawn. Like recreational usage, environmental usage is non-consumptive but may reduce the availability of water for other users at specific times and places. For example, water release from a reservoir to help fish spawn may not be available to farms upstream.

Water Stress

The concept of water stress is relatively simple: According to the World Business Council for Sustainable Development, it applies to situations where there is not enough water for all uses, whether agricultural, industrial or domestic. Defining thresholds for stress in terms of available water per capita is more complex, however, entailing assumptions about water use and its efficiency. Nevertheless, it has been proposed that when annual per capita renewable freshwater availability is less than 1,700 cubic meters, countries begin to experience periodic or regular water stress. Below 1,000 cubic meters, water scarcity begins to hamper economic development and human health and well-being.

Population Growth

In 2000, the world population was 6.2 billion. The UN estimates that by 2050 there will be an additional 3 billion people with most of the growth in developing countries that already suffer water stress. Thus, water demand will increase unless there are corresponding increases in water conservation and recycling of this vital resource.

Increased Affluence

The rate of poverty alleviation is increasing especially within the two population giants of China and India. However, increasing affluence inevitably means more water consumption: from needing clean fresh water 24 hours a day, 7 days a week and basic sanitation service, to demanding water for gardens and car washing, to wanting jacuzzis or private swimming pools.

Expansion of Business Activity

Business activity ranging from industrialization to services such as tourism and entertainment continues to expand rapidly. This expansion requires increased water services including both supply and sanitation, which can lead to more pressure on water resources and natural ecosystems.

Rapid Urbanization

The trend towards urbanization is accelerating. Small private wells and septic tanks that work well in low-density communities are not feasible within high-density urban areas. Urbanization requires significant investment in water infrastructure in order to deliver water to individuals and to process the concentrations of wastewater – both from individuals and from business.

These polluted and contaminated waters must be treated or they pose unacceptable public health risks. In 60% of European cities with more than 100,000 people, groundwater is being used at a faster rate than it can be replenished. Even if some water remains available, it costs more and more to capture it.

Climate Change

Climate change could have significant impacts on water resources around the world because of the close connections between the climate and hydrologic cycle. Rising temperatures will increase evaporation and lead to increases in precipitation, though there will be regional variations in rainfall. Overall, the global supply of freshwater will increase.

Both droughts and floods may become more frequent in different regions at different times, and dramatic changes in snowfall and snowmelt are expected in mountainous areas. Higher temperatures will also affect water quality in ways that are not well understood. Possible impacts include increased eutrophication. Climate change could also mean an increase in demand for farm irrigation, garden sprinklers, and perhaps even swimming pools.

Depletion of Aquifers

Due to the expanding human population, competition for water is growing such that many of the worlds major aquifers are becoming depleted. This is due both for direct human consumption as well as agricultural irrigation by groundwater. Millions of pumps of all sizes are currently extracting groundwater throughout the world.

Irrigation in dry areas such as northern China and India is supplied by groundwater, and is being extracted at an unsustainable rate. Cities that have experienced aquifer drops between 10 to 50 meters include Mexico City, Bangkok, Manila, Beijing, Madras and Shanghai.

POLLUTION AND WATER PROTECTION

Water pollution is one of the main concerns of the world today. The governments of many countries have striven to find solutions to reduce this problem. Many pollutants threaten water supplies, but the most widespread, especially in underdeveloped countries, is the discharge of raw sewage into natural waters; this method of sewage disposal is the most common method in underdeveloped countries, but also is prevalent in quasi-developed countries such as China, India and Iran.

Sewage, sludge, garbage, and even toxic pollutants are all dumped into the water. Even if sewage is treated, problems still arise. Treated sewage forms sludge, which may be placed in landfills, spread out on land, incinerated or dumped at sea. In addition to sewage, nonpoint source pollution such as agricultural runoff is a significant source of pollution in some parts of the world, along with urban stormwater runoff and chemical wastes dumped by industries and governments.

Water and Conflict

The only known example of an actual inter-state conflict over water took place between 2500 and 2350 BC between the Sumerian states of Lagash and Umma. Yet, despite the lack of evidence of international wars being fought over water alone, water has been the source of various conflicts throughout history. When water scarcity causes political tensions to arise, this is referred to as water stress. Water stress has led most often to conflicts at local and regional levels. Using a purely quantitative methodology, Thomas Homer-Dixon successfully correlated water scarcity and scarcity of available arable lands to an increased chance of violent conflict.

Water stress can also exacerbate conflicts and political tensions which are not directly caused by water. Gradual reductions over time in the quality and/or quantity of fresh water can add to the instability of a region by depleting the health of a population, obstructing economic development, and exacerbating larger conflicts. Conflicts and tensions over water are most likely to arise within national borders, in the downstream areas of distressed river basins.

Areas such as the lower regions of China's Yellow River or the Chao Phraya River in Thailand, for example, have already been experiencing water stress for several years. Additionally, certain arid countries which rely heavily on water for irrigation, such as China, India, Iran, and Pakistan, are particularly at risk of water-related conflicts. Political tensions, civil protest, and violence may also occur in reaction to water privatization. The Bolivian Water Wars of 2000 are a case in point.

World Water Supply and Distribution

Food and water are two basic human needs. However, global coverage figures from 2002 indicate that, of every 10 people:

- Roughly 5 have a connection to a piped water supply at home (in their dwelling, plot or yard).
- 3 make use of some other sort of improved water supply, such as a protected well or public standpipe.
- 2 are unserved.
- In addition, 4 out of every 10 people live without improved sanitation.

At Earth Summit 2002 governments approved a Plan of Action to:

- Halve by 2015 the proportion of people unable to reach or afford safe drinking water. The Global Water Supply and Sanitation Assessment 2000 Report (GWSSAR) defines "Reasonable access" to water as at least 20 liters per person per day from a source within one kilometre of the user's home.
- Halve the proportion of people without access to basic sanitation. The GWSSR defines "Basic sanitation" as private or shared but not public disposal systems that separate waste from human contact.

As the picture shows, in 2025, water shortages will be more prevalent among poorer countries where resources are limited and population growth is rapid, such as the Middle East, Africa, and parts of Asia. By 2025, large urban and peri-urban areas will require new infrastructure to provide safe water and adequate sanitation. This suggests growing conflicts with agricultural water users, who currently consume the majority of the water used by humans.

Generally speaking the more developed countries of North America, Europe and Russia will not see a serious threat to water supply by the year 2025, not only because of their relative wealth, but more importantly their populations will be better aligned with available water resources.

North Africa, the Middle East, South Africa and northern China will face very severe water shortages due to physical scarcity and a condition of overpopulation relative to their carrying capacity with respect to water supply.

Most of South America, Sub-Saharan Africa, Southern China and India will face water supply shortages by 2025; for these latter regions the causes of scarcity will be economic constraints to developing safe drinking water, as well as excessive population growth. 1.6 billion people have gained access to a safe water source since 1990. The proportion of people in developing countries with access to safe water is calculated to have improved from 30 per cent in 1970 to 71 per cent in 1990, 79 per cent in 2000 and 84 per cent in 2004. This trend is projected to continue.

ECONOMIC CONSIDERATIONS

Water supply and sanitation require a huge amount of capital investment in infrastructure such as pipe networks, pumping stations and water treatment works. It is estimated that Organisation for Economic Co-operation and Development (OECD) nations need to invest at least USD 200 billion per year to replace aging water infrastructure to guarantee supply, reduce leakage rates and protect water quality. International attention has focused upon the needs of the developing countries.

To meet the Millennium Development Goals targets of halving the proportion of the population lacking access to safe drinking water and basic sanitation by 2015, current annual investment on the order of USD 10 to USD 15 billion would need to be roughly doubled.

This does not include investments required for the maintenance of existing infrastructure. Once infrastructure is in place, operating water supply and sanitation systems entails significant ongoing costs to cover personnel, energy, chemicals, maintenance and other expenses. The sources of money to meet these capital and operational costs are essentially either user fees, public funds or some combination of the two.

But this is where the economics of water management start to become extremely complex as they intersect with social and broader economic policy. Such policy questions are beyond the scope of this chapter, which has concentrated on basic information about water availability and water use. They are, nevertheless, highly relevant to understanding how critical water issues will affect business and industry in terms of both risks and opportunities.

BUSINESS RESPONSE

The World Business Council for Sustainable Development in its H2OScenarios engaged in a scenario building process to:

- Clarify and enhance understanding by business of the key issues and drivers of change related to water.
- Promote mutual understanding between the business community and non-business stakeholders on water management issues.
- Support effective business action as part of the solution to sustainable water management.

It concluded that:

- Business cannot survive in a society that thirsts.
- You don't have to be in the water business to have a water crisis.
- Business is part of the solution, and its potential is driven by its engagement.
- Growing water issues and complexity will drive up costs.

WATERSHED ATLAS OF INDIA

The watersheds are natural hydrological entities that cover a specific aerial expanse of land surface from which the rainfall runoff flows to a defined drain, channel, stream or river at any particular point. The terms region, basin, catchment, watershed etc are widely used to denote hydrological units. Even though these terms have similar meanings in popular sense, technically they are different. Size of a watershed is governed by the size of the stream occupied by it. Size of the watershed is of practical importance in development programmes.

For example, size of irrigation cum hydel project has its watershed size several thousands of square kilometers but for a farm pond the size may be few hectares only. In deserts and flat terrains with little incipient drainage, it may be difficult to delineate small sized watersheds whereas in undulating and hilly terrains smaller sized watersheds could be easily delineated. Hence the aerial extent of watersheds vary widely in the various attempts made earlier for demarcation of watersheds.

A systematic delineation of river basins was first attempted in 1949 by Central Water and Power Corporation (CWPC) under the able leadership of Dr.A.N.Khosla, in which entire country has been distinctly delineated into 6 Water Resources Regions (Watershed Atlas of India, AISLUS, 1990). The 6 Water Resources Regions has been further divided into 66 major river catchments. These delineations have been extensively used in the planning and development of surface water resources of the country. Subsequently it was realized that there is a need for systematic delineation of the rivers systems at a national level on a convenient working scale.

In this context, Ministry of Agriculture, Government of India had assigned the work of preparation and codification of River Systems of India to Mansinghal consultants. In 1970, the Mansinghal Consultants suggested / published the methodology based on the area concept. The suggested hydrological units are: Water resource region, river basins, river sub-basin, watershed, sub watershed, mini-watershed and micro watershed. The recommended size of the watershed is from 250-750 Sq. Km.

Increasing thrust in water sector demanded finer details of hydrological units for implementation of various watershed management schemes. One of the major landmarks in this direction was by the All India Soil and Land Use Survey (AISLUS) Organization of the Dept. of Agriculture and Cooperation by publishing the National level watershed atlas on 1: 1 million scale using the base map from Irrigation Atlas of India in the year 1990. In this Atlas, the entire river systems of the country have been divided into 6 Water Resources Region, which has been further, divided into 35 basins and 112 catchments.

These catchments have been further divided into 500 sub-catchments and 3237 watersheds. The atlas consists of 17 sheets on 1:1 million scales along with a Compendium of watersheds giving details of other related information such as area within the basin, sharing states and stream names etc. This atlas is being extensively used for various purposes.

The attempts made so far for delineation of hydrological boundaries up to watershed level has some limitations. The atlas giving details of watershed is available on 1 million scale. However, for effective implementation of water resources development scheme there is growing requirement of a larger scale map. Further, these maps are available in hard copy and require digitization for integration in GIS platform. The terminologies and codification system adopted for watershed is also cumbersome.

NEED FOR LARGE SCALE WATERSHED MAPS

In the last two decades, watershed management has gained the top most priority in water resources sector. Implementation of any water management measure requires a suitable hydrological unit. A properly delineated watershed forms a convenient hydrological unit for computation of water balance parameters and thus implementation of water management schemes. The watershed approach has become a pre requisite for any developmental programme, because land and water resources have maximum interaction and synergic effect, when developed on watershed basis.

Watershed approach is, therefore, increasingly applied in various development programmes like command area development, soil and water conservation, flood control, soil erosion control, river valley projects, land reclamation, people and resource dynamics etc. It is also equally important for various hydro-power and irrigation projects, assessment of ground water resource, pollution and artificial recharge studies. For proper planning and execution of any development programme on watershed basis, it is essential to have watersheds in the form of maps along with relevant attributes. Increasing application of GIS techniques in water sector requires digital maps and geodatabase.

Under the Hydrology Project Phase – I, CGWB had the mandate to prepare the GIS data set on 1:250, 000 scale for various thematic layers relevant to groundwater for integration in the dedicated software and consequently in the Hydrological Information System (HIS) developed in the project. Watershed was one of the layers envisaged for integration in the software to facilitate the Ground water assessment on watershed basis. CGWB was identified by Ministry of Defence as one of the nodal agency for creation of digital data under the Hydrology Project.

Keeping in view the growing requirements of watershed delineation on 1:250,000 scale and its utilities in various departments for implementing diverse developmental activities an urgent need was felt to delineate the watershed boundary on 1: 250,000 and bring the data in GIS platform which can be utilized by different departments/ agencies for various purpose. The present Watershed Atlas is the outcome of the effort made by CGWB in this direction.

Chapter 2

Tank Irrigation Systems

Tank irrigation contributes significantly to agricultural production in parts of South and Southeast Asia. Especially in South India and Sri Lanka, tank irrigation has a long history and many currently used tanks were constructed in the past centuries. The tanks have existed in India from time immemorial, and have been an important source of irrigation especially in southern India. They account for more than one-third of the total irrigated area in Andhra Pradesh, Karnataka and Tamil Nadu states.

The tank irrigation system has a special significance to the marginal and small scale farmers who make a very large number essentially depending on tank irrigation as these systems are less capital-intensive and have wider geographical distribution than large projects. An irrigation tank is a small reservoir constructed across the slope of a valley to catch and store water during rainy season and use it for irrigation during dry season.

Tank irrigation systems also act as an alternative to pump projects, where energy availability, energy cost or ground-water supplies are constraints for pumping. The distribution of tanks was quite dense in some areas. However, over years the performance of the tanks has been declining. The share of tank irrigated area in India has declined from 16.51 per cent in 1952-53 to 5.18 per cent in 1999-2000, whereas the share of groundwater irrigation has increased from 30.17 per cent to 55.36 per cent during this period.

The share of the tank irrigated area to net irrigated area (NIA) had been declining continuously over years. Among the three major sources of irrigation, tank is the only source, where the irrigated area has been declining continuously since early seventies and many argue that the area under tank irrigation started declining only after the introduction of the green revolution.

Further, among the states in India, the area under tank irrigation has declined more drastically in those states where tank irrigated area accounts for relatively a larger share in the net irrigated area and it has increased marginally in certain states where it accounts for very low share in the net irrigated area. Data from the Agricultural Census of India for five time points namely 1970-71, 1976-77, 1980-81, 1985-86 and 1990-99 indicated that the resource poor farmers (owning less than 2 hectares) still account for major share of tank-irrigated area in India.

Marginal (less than 1 ha.) and small farmers (1-2 ha) together accounted for about 40 per cent of tank-irrigated area in 1970-71, which further increased to nearly 55 per cent in 1990-91 thus accounting for nearly two third of tank irrigated area. On the other hand, the share of tank irrigated area used by large farmers declined from 13.59 per cent to 6.02 per cent during this period.

Since the farmers belonging to marginal and small size group are mostly poor, they couldn't afford for cost-intensive irrigation sources like groundwater as in the case of medium and large farmers and tank irrigation continues to play a crucial role among small and marginal farmers even today. This is also true across different states where tank irrigation has considerable presence even today.

MANAGEMENT OF TANKS

In ancient days, tanks were considered to be the property of rulers. The farmers paid a portion of the produce to the ruler. Farmers also were in charge of the maintenance of the tanks, and supply channels. Zamindars ensured the proper maintenance of the tanks, and channels, since they reaped the benefits of farming in large areas. However, when the British introduced the ryotwari system in 1886, tanks with an ayacut of 40 ha. and above were brought under the control of Public Works Department (PWD) and smaller tanks were under the administrative control of local bodies, or vested with the villagers themselves.

Since the local bodies did not have qualified engineers and the duties of the ayacutdars were not clearly mentioned, the system of the farmers themselves taking up maintenance work known as kudimaramathu works slowly declined. Tanks were silted up, and supply and distribution channels choked. The deterioration of the tank irrigation system has been a subject of considerable discussion, at least since the middle of the 19th century.

The Report of the Public Works Commission of 1852 stated that there was not much of voluntary community labour involved in tank maintenance, and it reported that in all districts labour was more or less forced to work. In fact an act was passed, namely the Madras Compulsory Labour Act of 1858 (or what is known as the Kudimaramath Act), with a view to legalising compulsory labour for certain aspects of maintenance, and also to penalize the non-performance of kudimaramath labour. The entire administration of the act of levying and collection of fines was left with the irrigation panchayats. The Famine Commission of 1878 brought to light quite forcefully the deteriorating conditions of tanks and advocated a systematic policy of maintenance. However, at present the local village is responsible for water distribution and management of the tanks with a command area of below 40 ha.

PERFORMANCE OF TANKS OVER YEARS

In a 10-year period, three years the tank get normal supply, five-years get deficit supply and two years they fail completely. Given the rainfall uncertainties, the tank performance is seen declining over years. There are above-outlet problems such as poorly maintained structures (bunds, surplus weirs). Catchment is mismanaged and forest land adjacent to the catchment are already converted for human settlement by the Government. There are severe encroachment in the tank foreshores. Siltation of the tankbed has reduced the water storage capacity ranging from 20 to 30 per cent. In the case of below-outlet problems, channels are not maintained and broken resulting in heavy water losses.

Well irrigation has dominated the tank irrigation in several cases where the increase in the number of wells in the tank command had been signaling the inactiveness of the tank systems for providing reliable water supply. In fact, it had been found that large number of tanks have become defunct in less tank intensive districts (i.e., 76% of Panchayat Union tanks and 64% of Public Works Department tanks have become defunct) compared to tank intensive regions, where the percentage of defunct tanks is less.

Compared to all India level, in Tamilnadu state, the share of tank irrigated area to net irrigated area by marginal farmers has decreased from 39.53 per cent in 1970-71 to 35.17 per cent in 1990-91, tank irrigated area by small farmers had decreased from 32.02 to 0.23 per cent, medium farmers (2-4 ha) has decreased from 30.03 per cent to 21.47 per cent and large farmers (more than 4 ha) has decreased from 28.46 per cent to 19.40 per cent during the above periods respectively indicating the poor performance of the tank irrigation systems in the state.

The neglect of tanks has meant that most farmers receive inadequate quantities of water from tanks. To offset the decline in tank water supplies, farmers have resorted to supplemental well irrigation to avoid crop losses. Since only about 15 % of the farmers in the tank command area own wells and there is a growing demand for well water, the well owners in most cases act like local monopolists and are able to charge high prices for well water.

However, profit-making through privately owned water source (i.e. wells) within the hydrological boundary of the common property resource (tanks) poses serious threat to the very survival of the tanks, because of the declining interest among well-owners in proper upkeep of tank structures.

Even though several factors have influenced the tank performance, the level of their influences was varying across locations. The well density had negative influence of the tank performance. It was observed that higher the well density, lesser was the tank performance. Tanks without the well supplementation in the tank season had performed well and this clearly indicated the availability of adequate tank water supplies.

Concerning the O & M expenditure on tanks at state-level, the results of the study had indicated that though outlay per hectare of command area at current prices increased from Rs.26 to Rs. 161 per ha, the outlay at constant (1980-81) prices has increased only marginally from Rs. 33 to Rs.43 per ha. However, the level of O & M amount spent on the sample tanks revealed that the average amount spent was high for PU tanks (Rs 154/ha) compared to PWD tanks (Rs 74/ha). Since the O&M amount was spent mainly depending upon the urgency of the tank repair and the local political pressure, the level of tank performance and the amount of O&M spent could not be directly related.

MULTI-USES FROM TANKS

Even though tanks were originally serving irrigation and other village needs such as domestic, livestock, besides fish production, due to change in the village profile over years, tanks are mostly saving the irrigation needs only. However, judging the tank performance using the irrigation component may be inadequate, as it will not reflect the true performance of the tank benefiting the village in several ways.

Hence, multi-uses of the tank should be considered in arriving the tank performance. If such uses are in reasonable proportion, then rethinking tank management in terms of multi-use performance may be warranted. Also using the multiple benefits approach will indicate the magnitude of the receipts from all the uses which can be effectively used for tank maintenance.

In absolute terms, social forestry raises the most revenue (averaging Rs 170/ha), followed by irrigation (Rs 88/ha) and fisheries (Rs 15/ha). Social forestry collects the highest revenue (100%) as a proportion of total value of output, but irrigation pays a relatively small proportion of the value of output (3.2 %) in various fees. Social forestry appears to perform well in absolute, as well as relative revenue realisation at the tank level. The State Revenue Department, Social Forestry Department, Mines Department, Panchayats, and informal organizations in the village community are all involved in collecting revenue from the tank users.

Among the various agencies, Panchayat Unions receive the maximum realised revenue (64.96%), followed by the Social Forestry Department (24.84%), village community (5.18%), and the Revenue Department (4.67%). But if the panchayats generate so much income from the tank uses, why are they not investing more in attending to the maintenance of the tanks?

The panchayats feel that it was the responsibility of the state government to pay for the maintenance, and therefore did not put their own resources into tank maintenance. It is not clear what effect the Panchayati Raj Amendment has had on this situation, but it is essential to explore what will happen if the responsibility for tank maintenance as well as the entire revenue collection authority is given to a single institution such as local panchayats or water users association.

It is important to note from the tables that the total revenue realized in terms of taxes, fee etc., ranges from Rs 337.12/ha in PU tanks to Rs 270.29/ha in PWD tanks, with an average realization at the tanks as Rs 275.40 /ha. This is higher than the government allotment of Rs 140/ha for tank O&M. Hence, instead of receiving heavy small allotments from the government, in fact tanks themselves can generate more resources for maintenance.

Present practices do not seem to be even exploiting the full potential of tapping all the uses of tanks for revenue to support them and hence tanks suffer from lack of maintenance funds which is one of the major reasons for poor condition of the tanks. However, further analysis is needed to determine whether the revenue generation will be uniform across tanks, and how different combinations of uses may be competitive and or complimentary in nature.

WARNING SIGNALS

The following are the warning signals to the Government and local community on the declining tank irrigation in the state:

- Mostly tanks are reported to be functioning only in normal and excess rainfall years and not so in poor and low rainfall years. The consequences are: many farmers have started abandoning tank agriculture due its continuous uncertainties in water supplies and moving to the nearby towns for other jobs and only the older people are staying back in the tank villages. The lands are not maintained properly and the prosophis trees are growing freely in the cultivated lands thus making the lands unsuitable for cultivation during years when the tank has adequate water. Due to the declining commitment on the maintenance of the tank structures, the upkeep of the structures is a cost affair for the farmers when they really want to use the tank for irrigation during normal supply periods.

- The livestock support activities are also completely gone in the village eco-system thus eroding the livelihood options in the village. Farmers used to take the silt using bullock carts and after the introduction of the social forestry scheme in the 1980s in the water spread area, silt removal from the tanks was prevented thus making the bullock operations limited. Somehow in the recent years, the micro-finance concept has been emerged among the rural women who are managing the families with livestock and credit integration. But livestock also need adequate fodder. Hence, if the tanks are not properly managed then the entire tank-ecosystem based rural economy will be completely collapsed.
- The impact of the social forestry was already felt in terms of increasing silt accumulation in the tank water spread area and it will be difficult to sustain the tanks if the social forestry is allowed to continue. But at the same time, even without social forestry in the tanks, there are possibilities that the prosophis trees will be spreading fast and it will have severe impact than the social forestry with accacia trees which have market (timber) value.
- It is seen in several locations, due to intensification of watershed development programs by the Government, several structures such as small check dams and percolation ponds are developed in the upstream of the tanks thus affecting the inflows into the tanks. Hence a clear demarcation should be done between the watershed programs and tank improvement programs.
- Disappearance of the supply channels is very common. House construction works due to population increases and village development activities such as roads, schools, buildings are concentrated in the government poramboke (common) lands which are the main sources of inflow to the tanks as well as

interlinking the tanks in the chain. This is one of the reasons why tanks are not getting adequate storages even though the rainfall is normal.

- The traditional village institutions like needkatti or madayan thotti who looked after the tank catchment and tank structures and facilitated the inflows into the tanks regularly during rainy seasons also disappeared, as they could not be paid by the farmers due to frequent tank failures.
- The growing nexus between castes and politics among the younger generation in the village also played their role in making the traditional leaders in the village (who looked after the tank management) inactive. Several regional political parties are coming up and since the voting percentage is higher in the villages, these parties concentrate on the rural villages for their benefits and in the process, the households are divided among the political and caste related groups.
- The growing self-interest and non-cooperation by the well owners in the routine tank maintenance also make the tank management a difficult task. This is because in several villages, well owners feel that the tanks will not be much useful, as most of the periods they are dry. Also the reliability of the tanks for recharging the wells has also gone down due to siltation and encroachment.
- The rice supplies in the village ration shops to some extent make the poor farmer households to prolong their livelihood with the dried-up tanks. But the major issue is how long the ration shops will sustain the villages and the tanks.
- Many people now raise the question: Do we really need the tank bund which makes 1:2 or 1: 4 water spread : command area?. The 1:2 ratio (i.e., for every one hectare of water spread, only 2 hectares of command area is available) is very attractive for making the rainfed tanks into rainfed land as there is not much differences between the tank irrigation and rainfed agriculture. This aspect is gaining important since in most of the time, the tanks are empty and people think about using the water spread area for rainfed cultivation due to its fertile silt.

POLICIES FOR IMPROVING TANK SYSTEMS INVESTMENT

Tank rehabilitation options that restore the original standards should be given priority. Desilting is an important option. However, it was observed, that in a 10 year cycle, only in 3 years, the tanks get full storage, five years deficit storage and in two years the tanks fail. ence, desilting the tank fully will not be economical, as the benefits due to desilting will be in three years only, where the tanks get full supply. Also disposal of the entire desilted material is difficult, as the fertile silt is found only in the top (0.4 metre) layer.

Therefore, full scale desilting may not be warranted. Considering the high cost of Rs. 120/ m of silt, partial desilting that helps to restore original (10%) dead storage could be attempted as part of tank rehabilitation options as this will help increase non-irrigation benefits of tank water particularly in the non-tank-irrigation season. Also recharging of wells will be improved. Partial desilting can be done nearer to the lower sluice as well as around the periphery of the tank water spread area.

Most of the tanks are not getting adequate water supply and the chain system of tanks has almost broken. Hence, there is an urgent need to revive the tank-chains through appropriate modernization strategies for improving the supply channels connecting different tanks. This highlights the need for taking up modernization works at chain-level i.e. by considering the entire hydrological boundary as a single unit rather than viewing individual tanks as separate entities for new investment. Community wells should be installed in the tank water spread area to provide few supplementary irrigations to the non-well farmers during critical periods.

MANAGEMENT

Farmers in few water scarcity tanks have already been adopting crop diversification strategy involving groundnut, pulses, cotton and other crops and this practice should be extended to tanks whose water storage is 50-60 per cent. The water required to produce one kilogram of rice ranges from 4500 - 5000 litres compared to 1500 - 2000 litres in the case of non-rice crops such as groundnut. Hence, using the 50 per cent tank storage, the entire command area can be covered with non-rice crops. Extension efforts and marketing support to farmers should be strengthened to introduce crop diversification particularly in the wet season. Crop demonstrations by the Department of Agriculture should help speed up the process.

To complement the above options, tank structures should be repaired for effective water control. Water losses in the canals are about 30 per cent besides creating inequity in distribution between head and tail farms. Lining the main canals can be followed without disturbing the field boundaries. Tank management strategies such as sluice rotation will help save the tank water by 20 per cent. Instead of continuous water withdrawal from tanks, sluices can be opened and closed on alternate weeks (rotation of sluices).

LEGAL

More tank have become defunct in the recent years due to encroachment, siltation, choking of supply channels and pollution from industries. Tanks close to the cities should be protected from environmental pollution and further be made as groundwater recharges structures for domestic purposes. Strict regulations and penalty mechanisms should be imposed on the encroachers of catchment, supply channel, and foreshore area.

IRRIGATION DEVELOPMENT, POLITICS AND SOCIETY

Initially, India's Watershed Development Programme (WDP) was an externally implemented solution aimed at addressing the problem of degraded lands and their resource poor inhabitants. Following a review in the early 1990s the ideology evolved towards a more community-based, participatory approach. The combination of a watershed as an area for development, and the involvement of local people, was considered to be so successful that other Government programmes with similar development objectives were grouped under the Guidelines for Watershed Development.

The failing Desert Development Programme (DDP) that was operating in both the hot Western deserts and the cold, high altitude deserts of Jammu and Kashmir State was one such programme that integrated this shift in approach. Thus, in 1996, the Watershed Development Programme was introduced through the DDP to selected communities in the arid, high altitude region of Ladakh.

A preliminary survey deemed Ladakh's Zangskar Valley to be unfit for the Watershed Programme at this time. However, in due course this remote mountainous area was reassessed and in 2001 the first concrete steps were taken towards WDP implementation. This study explores the continuing development of Zangskari irrigation through the agendas and perceptions of the parties involved in the implementation of the Watershed Development Programme. The analysis is carried out in the context of existing irrigation development and both historical and contemporary social, political and environmental factors.

The effectiveness of the Programme is also assessed with particular reference to the needs of the Zangskari farmer and the re-empowerment of village communities. The study draws upon secondary data and aspects of social and political theory, but the majority of the evidence comes from primary data collected during a field visit to Zangskar in June and July, 2003.

Plagued in recent years by failing irrigation development and fluctuations in the availability and reliability of irrigation water, the WDP theoretically offers Zangskaris a rare opportunity to improve their water security through community mobilisation and sustainable development. However, in reality the Programme in Zangskar is being steered by a complex set of agendas which in many instances deviate from the Watershed ideology and the agricultural requirements of the Zangskari farmer.

PHYSICAL GEOGRAPHY AND TENUOUS COMMUNICATIONS

Encompassing 7000 square kilometres of high altitude, arid terrain, Zangskar tehsil is sandwiched in southern Ladakh between the Greater Himalayan Range to the south west and the Zangskar and Ladakh Ranges to the north east. Situated on the northern fringe of the Western Himalaya, Zangskar, like the rest of Ladakh and the neighbouring Tibetan Plateau, is subject to the Himalayan rain shadow. Precipitation in Zangskar is largely confined to winter snowfall which effectively isolates the central valley and surrounding settlements from the rest of India for up to eight months of every year.

Climatic extremes reflect the fact that little of Zangskar is below 3500m and this in turn is reflected in an agricultural season which is compressed between early May and mid September. During the brief summer, travel in and out of the region is restricted to either a single 235km unsurfaced road or a number of trekking routes which negotiate one or more 5000m passes. An infrequent bus service and commercial vehicles make use of the road at this time, although travel to either Kashmir or the rest of Ladakh - which is via the disputed Indo-Pakistan border area around Kargil - is often disrupted by military action.

TRADITIONAL AGRICULTURE, LABOUR DEMANDS AND MODERN AGRICULTURAL TRENDS

Historically, agriculture has been the mainstay of the Ladakhi economy. Despite an increase in possible sources of income, primarily through government employment, military service and the growing tourism industry, the centrality of agriculture in Ladakh persists to this day. Dependence upon agriculture is particularly apparent in Zangskar where over 95% of people are estimated to still be reliant on the productivity of their own farms. The subsidised 'government rations' that are trucked into the Valley during the brief summer remain, for many people, an expensive luxury that supplement, rather than replace locally produced cereals, vegetables and dairy products.

For many Zangskaris, especially village elders, a successful harvest is a symbol of self-sufficiency, an important element of Zangskari identity that is still highly valued in this remote environment. Despite the harsh climate and limited growing season traditional Zangskari agriculture is remarkably productive with yields often being comparable to those obtained from European intensive regimes.

Achieving such high levels of productivity is partly due to a high standard of cultivation practice which ultimately demands high labour inputs. Labour requirements peak at harvest time and every available hand from the farming estate is employed for the busiest period of the agricultural calendar. In the broader picture of Zangskari work patterns, agriculture is one of many activities that demands time and labour during the warm summer months.

In recent years however, the introduction of additional time-intensive activities such as employment and schooling has made meeting these demands increasingly difficult. This decline in the available workforce is now being overcome primarily through the employment of labourers - a practice which indicates a shift towards the new cash economy.

Largely restricted to wealthier Zangskaris, the use of seasonal labour is a significant agricultural development in communities which still contain a high proportion of subsistence farmers. It is the prevalence of subsistence farming which, perceived by the government as being backward and unproductive, is motivating the majority of agricultural development in Zangskar. Modern chemical fertilisers and machinery such as threshers, tractors and water pumps are being offered to Zangskari farmers at subsidised rates under the pretext of boosting agricultural productivity.

In addition to this, 280,000 hybrid seeds and seedlings were recently distributed (spring 2003) at 100% subsidy with a view to improving nutrition levels and encouraging the cultivation of cash crops. This encouraged diversification of farming practice has also resulted in less productive, marginal land being allotted for willow plantations. Having recognised the benefits of a ready source of wood and fodder, Zangskari farmers have begun to manage large plantations at the village level and they have also allocated valuable land and water to this new initiative.

WATER RESOURCES AND TRADITIONAL IRRIGATION SYSTEMS

Having evolved under the influence of the Himalayan rain

shadow, Zangskari cultivation is entirely irrigation-dependent. Indeed, throughout Ladakh, water is widely recognised as being one of the most significant factors constraining the size of settlements and the area under cultivation. Although a number of factors influence whether a soil in Zangskar is cultivable, proximity to a reliable water source is critical. Villages are therefore positioned in order to utilise the melt-water from streams and springs which drain neighbouring glaciers and semi-permanent snowfields. Carefully engineered channels harvest this water and transport it, sometimes over several kilometres of rugged terrain, to the area under cultivation.

Villages are reliant upon small scale gravity fed channels as Zangskar's major rivers are largely unusable. Not only would spring floods in these rivers destroy the head works of the irrigation channels, but these high energy flows have also cut deep into the landscape and thus they are generally too low to feed into gravity fed irrigation systems. Influenced by a host of variables, the availability and reliability of irrigation water that is received alters from village to village.

Variables may include: the aspect and size of glaciers or snowfields; climate and weather fluctuations (especially the previous winter's snowfall); the degree of channel maintenance; topography and soil/rock type between the source and the village; and the nature and size of the area under cultivation. It is therefore no surprise that the amount of water a village receives "varies by year, season, day and even hour".

In response to this variability, and potential scarcity, the water allocation systems that exist in each village incorporate a degree of flexibility into the fixed framework which ensures equitable and efficient distribution. Operating in a rotational system, water-user groups may be organised according to field, irrigation channel or household.

Members of the user groups have the right to use water for an allotted period of each rotation and this 'water time' is informally regulated by a transparent system in which villagers monitor each other. Many irrigation channels in Zangskar are centuries old and the right to use water from these channels is usually inherited. Water allocation generally favours the households that claim 'full' ancestral rights and without such a claim it is very difficult to break into these closed village systems.

This 'club' mentality to water rights effectively discourages newcomers and the further division of village water resources. Such a policy also reflects the extent to which household power, productivity and wealth is linked to the securing and protection of water rights.

Not only is irrigation fundamental to life in Zangskar it is also woven into the social and political fabric of Zangskari villages.

As a communal resource, individuals have to seek the consent of the village before building new fields and channels that require water from the existing irrigation network. Under the organisation of the village headman (mog dam), the maintenance of village irrigation channels, or the less frequent construction of new channels, also reflects this communal approach as traditionally the labour for these tasks is provided by each household.

Even households that are unable to offer labour contribute by supplying food and drink for those working on village projects. Aside from the annual removal of silt, traditionally constructed irrigation channels generally require little structural maintenance. The roots of grasses and herbaceous plants that grow along the banks of the channel bind the soil and help maintain channel structure.

However, in very sandy soils, or where the irrigation channel crosses areas of scree, excess seepage is often a problem. Whilst many villages enjoy the relative water security of large glacial reserves, others may be reliant upon smaller snowfields and the regular accumulation of winter snowfall. Indeed, a recent spate of poor winter snowfall and resulting summer droughts graphically illustrated the vulnerability of some of Zangskar's settlements.

Villages such as Karsha, Pishu and Zangla, proved to be particularly susceptible, and these and many other settlements throughout Zangskar employed drought strategies which involved decreasing, sometimes by up to two thirds, the area under cultivation. Kumi, a village that has experienced water shortages over the last 20 years, also suffered and activity is now underway to move the entire village down to the banks of the Lungnak River in an attempt to regain water security.

SETTLEMENT, HOUSEHOLDS AND POLYANDRY

According to the Indian census Zangskar consists of only 25 villages. This figure is however somewhat misleading as it refers to administrative units rather than single settlements. For example: an administrative 'village' may consist of a large settlement of some 70 or more houses, and several smaller contiguous 'hamlets' which, despite being separated from the main settlement by large expanses of unimproved land, fall under one census entry. In reality Zangskar has over 100 individual settlements although some may only contain two or three houses. Despite being distributed over a vast and diverse landscape, Zangskar's villages share common characteristics such as their relatively low altitude and their proximity to suitable water and cultivable soil.

The favourable convergence of these variables is however rare, not only in Zangskar but across much of Ladakh. Productive land is therefore highly prized and rules of inheritance ensure that the household, complete with farming estate, is passed down in its entirety to the next generation. As Gutschow points out "the Zangskari household is the principal nexus around which the economic relations of labour, production and subsistence are arranged".

Continuity of the household is of primary importance and this is traditionally facilitated in Zangskar's Buddhist community through the practice of polyandrous marriage and primogeniture. At some point after marriage the family's eldest son becomes the head of the household and inherits the 'main house' (khang pa or khang chen) and accompanying land.

The eldest son also inherits the responsibility of managing household affairs and the farming estate, often with the assistance of one or more of his brothers who may also share his wife. Aged parents move to a minor house (khang chung) where, depending on the labour requirements of the khang pa, they may also be joined by unmarried siblings. In the event that there is no reproductive brother to inherit the khang pa then the responsibility of heading the household falls to the eldest daughter who engages in a matrilocal marriage. As Crook points out whichever system of marriage is settled for, and it may simply be a monogamous arrangement, the continuity of the household and the entire farming estate is maintained within the same patrilineage.

Whilst polyandry maintains household unity it is also considered, in a more functionalist interpretation, to by a practice which inhibits population growth and therefore reduces the demands of a large population on a resource poor environment. The effects of these Buddhist practices are particularly apparent in the village of Padum where, in contrast, the rules of inheritance practised by Zangskar's Muslim minority have resulted in the equal division of farming estates between siblings.

In an environment where water and cultivable land is limited this practice eventually leads to the continued fractionalisation of land and progressively smaller plots which, as Singh points out, are not agriculturally viable. In Padum the Muslim community is now expressing concern over the increasingly small plots of land on which they have to farm. However, relatively recent developments have influenced this situation such that these concerns are no longer restricted to the Muslim community.

Polyandrous marriage and primogeniture was outlawed in Ladakh by State legislation in the early 1940s. Although the effects of this were not instantaneous, especially in remote regions such as Zangskar, a slow trend away from polyandrous marriage and primogeniture was initiated. In Zangskar's Buddhist community this shift is particularly apparent in Padum and neighbouring settlements, where the road link to Kargil and the opportunity of government employment have fuelled modernisation processes and the growth of a cash economy.

Younger (Buddhist) sons who traditionally would not qualify for inheritance rights have claimed their share of land and the result, especially in and around Padum, has been a sudden rise in the number of minor households (khang chung). The new cash economy has provided a means of overcoming any resulting shortfall in agricultural productivity or available labour. However, once outside Padum's economic sphere of influence, village growth is still restricted by the availability of cultivable land and, more importantly, water.

Although not conclusively linked, the decline of polyandry and primogeniture and the increase in monogamous nuclear households may at least in part be responsible for the population increase in Zangskar. Figures from the Revenue Office and Sub-Divisional Magistrate in Padum confirm a marked population growth over the last 30 years from 6,865 in 1973 to 12,167 in 2003. Yet, despite this rapid increase, Zangskar continues to be one of the least populated areas in Kargil District.

ADMINISTRATION, DEVELOPMENT, BUDDHISTS AND MUSLIMS

Zangskar is categorised as a tehsil, a Sub-Division and a (Community Development) Block within Kargil District, one of two Districts - the other being Leh - that constitutes the Ladakh region. Kargil District contains two tehsils, Kargil and Zangskar, and these in turn are divided into Blocks. Whilst Kargil tehsil is divided into six Blocks, Zangskar tehsil contains only one as the relatively small population does not warrant further division.

As development funds are allocated on a Block level, Zangskaris often consider themselves to be at a disadvantage, especially as Zangskar tehsil constitutes around 60% of the District's total land area. However, the designation of Sub-Division does provide Zangskar with an additional administration intended to boost the development of remote and 'backward' areas. As a Sub-Division Zangskar benefits from a block level administration headed by a Sub-Divisional Magistrate (SDM) and a Public Works Department (PWD) headed by an Executive Engineer.

The majority of development initiatives which are implemented in Zangskar originate at State or Central Government level. Within the scope of these government initiatives, funding applications for Block-level development are submitted by Zangskar's administration to the District administration based in Kargil. Subject to District-level approval, J & K State or Central Government is then approached for funding. If successful, funding is then released back through the system to the Zangskar Block. In theory, this administrative system represents a simple and efficient hierarchy.

In practice, even government employees admit that the percolation of directives and funds through this system is often slow and haphazard. Zangskar, like the rest of Ladakh, is well integrated into both Central and J & K State rural development initiatives. However, from the perspective of the Zangskari, development activity frequently falls short of addressing such fundamental issues as health care, education, agricultural development and water security.

Logistical problems associated with Zangskar's physical remoteness are undoubtedly responsible for part of this perceived neglect, however, the most frequently voiced reason, especially from Zangskar's Buddhist majority, is the division between Buddhist and Muslim communities. There is a lengthy historical and political background to Muslim and Buddhist relations in Ladakh.

However, a brief example illustrates the nature of the grievances that perpetuate what essentially is a divide based on the politicisation of religion. Prior to the bifurcation of Ladakh District in 1979 and the creation of the Buddhist majority Leh District and Muslim majority Kargil District, all State and Central Government Offices were centred in the predominantly Buddhist town of Leh. Under this Buddhist majority administration, the Muslim minority considered themselves to be disadvantaged by discriminatory policies that, even today, are considered to be the root cause of Kargil District's backwardness.

Regardless of whether the accusation is founded or not 'evidence' of this nature is continually recycled in the struggle for political representation and dominance. Thus the perpetuation of distrust, resentment and religious division is assured.

Far from suggesting that the current Kargil administration is employing a policy of retaliation against the Buddhist minority in Zangskar, the stage has been set for the less than cordial relations to persist. Indeed, Zangskar's Buddhists frequently accuse Kargil's Muslim majority administration of discrimination, a lack of political representation and the withholding of development funds and resources.

This division and general distrust is intensified further by an influential Muslim minority in Zangskar, who not only constitute the majority in the village of Padum, Zangskar's administrative capital, but also command a greater portion of government positions (and salaries) in Zangskar's Sub- Division administration. Events in the summer of 2003 have served only to perpetuate this situation. Kargil District has recently been granted Autonomous Hill Development Council (AHDC) status by the Jammu and Kashmir State Government.

Zangskar however has only been awarded 3 out of 26 seats on the Council and vocal protests by several (Buddhist) representatives from Zangskar have been largely ignored. In protest the Buddhist dominated Zangskar Action Committee intended to boycott the Hill Council elections in July 2003 and begin 'agitating' the Muslim administration. The implications of Zangskar's geographical and political isolation have also been discussed. Building upon this foundation this study will demonstrate how these factors are influencing the continued development of Zangskar's irrigation water through the implementation of the Watershed Development Programme.

THE DEVELOPMENT OF ZANGSKAR'S WATER RESOURCES

This chapter charts the brief history of Zangskar's water resource development and outlines the ideology and recent implementation of the Watershed Development Programme in Central Zangskar. Government departments In Zangskar there are three principal departments responsible for development of water resources, the Public Works Department (PWD), the Department of Rural Development (DRD) and the District Rural Development Agency (DRDA). Both the PWD and the DRD have a permanent presence in Central Zangskar with dedicated office complexes housing engineers and the Block Development Officer (BDO), the DRD's Block-level representative. The DRDA is also actively involved, although being a subsidiary organisation of the DRD it has less of a 'physical' presence. The DRDA's primary role is to oversee the District-level implementation of anti-poverty development programmes. The Agency operates as a catalyst organisation between the Ministry of Rural Development and various facilitating organisations at District and Block level where development programmes are implemented. In Zangskar, DRDA activity is overseen by the Kargil-based Project Officer DRDA and the District Development Commissioner.

IRRIGATION EFFICIENCY

Although the Zangskaris have built and maintained fine-tuned irrigation networks over generations, the arrival of the BDO in the 1950s and the PWD in the 1970s has lead to the integration of both of these government departments into village irrigation activities.

These departments are not restricted purely to irrigation development, however, such work falls well within their scope, and due to the centrality of irrigation in Zangskar a high percentage of BDO and PWD resources are directed into this area of development. Government assistance is available for the maintenance and improvement of channels or for larger construction projects. However, whilst Zangskaris have made full use of this potential source of labour, materials and funding, the actual planning and execution of work is often beyond their control.

The PWD frequently employ Zangskari or Kargili contractors and the BDO - despite having the means to employ villagers to work on their own projects - regularly opts for the employment of contracted labour. Input from what is perceived to be the 'backward' village often fails to penetrate India's social and political barriers. As a consequence village needs are not always fully explored.

To date, work conducted or funded by the PWD and BDO in Zangskar's irrigation sector has focused on improving productive efficiency.

In many instances the development of village irrigation networks has focused on 'modernising' traditionally constructed channels and reservoirs with concrete and cement linings. New irrigation channels have also been constructed out of cement lined dry stone walls. Many projects are on a relatively small scale although there are several large projects at various stages of completion.

Perhaps the two greatest icons of government involvement in Zangskar's irrigation development are the Haftal and Zangla canals. At an estimated cost of 24 lakh Rupees the PWD's 10 year Zangla Canal project was finally commissioned in the autumn of 2000. The three kilometre canal, one third of which is rather impressively cut into a cliff face, transports water from a previously unusable stream to Zangla's fields. The Canal has been so successful that Zangla's water allocation scheme has been renegotiated by the village in response to this new found water security.

In contrast, the Haftal Canal, which remains inoperative and in a state of disrepair, is arguably more symbolic of Zangskar's irrigation development. A product of the Desert Development Programme - a DRD initiative implemented through the DRDA aimed at mitigating the effects of desertification - the 15 year old

Canal has yet to irrigate the largely uncultivated floodplain south of the confluence of the Stod and Lungnak rivers.

FAILING IRRIGATION DEVELOPMENT

Despite the combined attention of the BDO, the PWD and the DRDA, much of the work to improve Central Zangskar's irrigation efficiency is either in a state of disrepair or unfinished. Sendo, a hamlet of Karsha, has a collection of metal pipes lying unused in a streambed simply because of a lack of pipe couplings. Shillingskit has a dry masonry reservoir that has not held water since its construction 10 years ago, and a new irrigation channel in Sani only reached a kilometre in length before it was abandoned.

A concrete channel above the village of Kumi failed to address their water deficit and in Pishu a new 16km canal was worked on for one season before it too was abandoned. Concrete and cement lined irrigation channels can be found in many villages, although these are frequently silted up, cracked and in a general state of disrepair.

Informants from the village-level attributed the poor success rate of the BDO, PWD and DRDA irrigation projects to corruption and selfishness within the system. The distrust was not only directed at the Block and District-level administration but also at Zangskari and Kargili contractors. Informants frequently stated that as little as one third of funds, released on the basis of a budgeted application, were invested in the proposed project.

Additionally, the workmanship of contracted labour was at times thought to be substandard. Contractors suggested that they had little choice but to appropriate a significant share of the project fund otherwise corruption at the departmental-level would result in them making a loss. Zangskar's Block administration however directed accusations back at the village-level where funds from the BDO, which again are released for specific budgeted projects, were often considered to be channelled into other unrelated village activities, or into individual pockets.

The state of disrepair observed in many concrete channels was also attributed by the Block administration to a lack of village-organised maintenance. However, at the village-level maintaining government built channels was almost always considered to be a government responsibility. The current problems surrounding Zangskar's existing irrigation development would appear to stem from the regular shortfall of funds and poor workmanship by people who are not necessarily the beneficiaries.

Furthermore, the high incidence of disrepair, specifically associated with government built channels and reservoirs, also implies discrepancies over ownership and a shift from traditional maintenance practices towards a partial dependence on external funds, materials and labour.

Indeed, Dawa (1999: 376) recognises this shift in Ladakh and notes the ability of subsidies and handouts to erode the characteristic qualities of self-reliance and sustainability. The recent drought conditions in Zangskar have only served to amplify the shortcomings of existing irrigation development, and as a consequence Zangskaris are now looking to the newly introduced Watershed Development Programme for solutions.

THE WATERSHED DEVELOPMENT PROGRAMME

Headed by Hanumantha Rao, a 1994 review of the Desert Development Programme and the Drought Prone Areas Programme recognised the need for the increased participation of user groups in watershed development. In response, the Ministry of Rural Development formulated the Watershed Guidelines in 1995 which encapsulated the participatory approach to improving India's degraded land. Using a watershed as a unit for development, the resulting Watershed Development Programme (WDPs) focused on the broad objectives of improving productivity and the production potential of ecologically fragile environments and their disadvantaged inhabitants.

With the advent of the Department of Land Resources, these objectives were further refined to enhance the viability and quality of rural livelihood support systems. Applied over a period of five years, the programme - which is funded by State (25%) and Central Government (75%) - employs sustainable technologies and the meaningful participation of beneficiaries to promote economic development and the improvement of socio-economic conditions.

Encouraged by the apparent transparency of funding - which is directly received by the communities involved - the participation of beneficiaries not only refers to project planning 32 and decision making but also their employment in the project construction. Participation not only boosts the local economy, it is also intended to contribute to a sense of ownership and obligation which encourages asset maintenance after project completion.

One of the key watershed objectives is that projects are sustainable, and as such, all maintenance activity is funded and carried out by the community that inherits the assets. Although the Watershed Guidelines have been updated since their conception in 1995 the ideology remains fundamentally the same and the success of the WDP is such that many of India's rural development initiatives have now been brought under the Guidelines.

Introduced to Zangskar in the autumn of 2001 the WDP's implementation has been through the DRDA administered Desert Development Programme (DDP) and the associated Desert Development Agency (DDA). However, at Zangskar's village-level, watershed projects are implemented by a Watershed Association (WA) and a Watershed Committee (WC).

Supervision and guidance of these Associations and Committees is provided in Zangskar by selected Non-Governmental Organisations (NGOs) which assume the role of Project Implementation Agencies (PIAs). Currently there are three Zangskari PIAs each working with approximately 10 village projects. PIAs are chosen - by the DRDA - on the basis that they have been active in related fields of development for some years. Receiving 20% of the total project fund (currently 30 lakh rupees) the PIA is responsible for site selection and a wide range of guidance from entry point activities, technical assistance and training, to Watershed Project accounts and the preparation of plans and post-project institutional arrangements for maintenance and further development.

To assist in these village-based activities the PIA employs a Watershed Development Team consisting of an appropriately qualified engineer, vet, agriculturalist and social scientist. Watershed Development Projects On the surface, all of the WDP 'building blocks' are in place in Zangskar for a smooth and successful implementation. Villages and PIAs have been paired and in many instances Watershed Projects are in the initial stages of construction.

Although Watershed Guidelines do not dictate the specific nature of projects, all but one of the selected villages have included irrigation development in their respective plans 37. This initial focus on irrigation is considered by PIAs to be an essential foundation upon which other more socio-economic developments can be built in the future. This approach was echoed in many villages but best summarised by an informant from Reru who claimed "if we have enough [irrigation] water then anything is possible".

THE 'ACTUAL' IMPLEMENTATION AND EFFECTIVENESS

Although at this early stage of construction it is not possible to comment on the ability of the projects to deliver irrigation water, the agendas and perceptions which ultimately influence the success of the WDPs are already becoming established. The effectiveness of the Programme's holistic ideology is also assessed from the perspective of the Zangskari farmer who, theoretically, is the main beneficiary.

Key Players in Zangskar's WDPs

It is useful here to draw upon Mary Douglas' Group Grid Theory which provides a useful means of identifying major groups of players and their relationships with each other. As Allan (2001: 320) points out the theory is not predictive "but it does help explain policy and operational outcomes".

When discussing the need for Watershed Projects, Zangskaris persistently stressed the importance of remaining largely self-sufficient; at least in terms of water and food security. An increasing reliance on transported goods and the tenuous road link with Kargil and the rest of India was certainly considered by many as being undesirable. Indeed, in the light of such extreme geographical isolation and the political instability of the neighbouring Indo-Pakistan border, both Osmaston (1989: 8) and Dawa (1999: 376) recognise the folly of such dependence. Buddhist Zangskaris also referred to the discriminatory policies of Kargil's Muslim administration as further strengthening their resolve to remain independent.

From an agricultural perspective the desire for self- reliance may therefore be viewed as a factor which fuels Zangskar's irrigation maintenance and development. A reliable and well maintained irrigation system is a tangible symbol of self-reliance and the continuity of Zangskari villages and households. When asked, the majority of informants from communities in Central Zangskar suggested a perceived compatibility between the scope of the WDP; the preservation of their self- sufficiency; and the continuity of settlements through the development and maintenance of this symbolic resource.

Despite originating at the village-level, these perceptions have certainly been reinforced by the PIAs who, acting as catalysts to WDP implementation, have promised that village participation and co-operation will secure funding, future water security, and a greater independence from Kargil's District administration.

Placed also in the context of existing irrigation development, Zangskari farmers - many of whom are employed in Watershed Projects - were excited by the perceived transparency of the system, the re-empowerment to make decisions and the fact that funds and materials bypassed contractors, the PWD and the BDO, by being delivered directly to the village- level. Despite this optimism, elders, and those with direct experience of BDO and PWD activities, were cautious of PIAs and the apparent transparency, implying that it was too early to comment and trust in the WDP would have to be earned. In the context of recent drought events the arrival of the WDP is also perceived by many Zangskaris as being rather timely.

Although irrigation distribution methods recognise the variability of their water supply, several successive summers of water scarcity have certainly highlighted the vulnerability of irrigation-dependent villages. The hardship associated with drought and the resulting increase in the awareness of risk is an example of what Douglas (1975: 8) refers to as the fore-grounding of information.

This prioritisation of perceived risk not only offers a further explanation for the dominance of irrigation-centred Watershed Projects, it also explains the urgency with which certain projects, particularly those in the water scarce villages of Kumi and Pishu, are being approached. Concerns over the erosion of Zangskari self-sufficiency suggest a realisation in Zangskar that both external and internal processes are threatening the sustainability of livelihoods.

In Central Zangskar, the decline of polyandry and primogeniture is one such process which is identified as being the root cause of problems surrounding population growth, the demand for village and agricultural expansion, and the increasing pressure on limited resources, especially water. Whilst the commissioning of the new canal has addressed this problem in the village of Zangla, concern was still very apparent in the settlements of Reru, Mune and Pishu. In these villages the drive behind WDP implementation was largely based on the potential to expand the village irrigation network and bring new land under cultivation.

The growth of these villages is currently restricted by a limited water supply and the reluctance of existing water users to either share their water rights or embark unaided on the massive task of building a new channel. In one village, which now only contains one polyandrous household, perceptions of the problem were so acute that potentially cultivable wasteland was already being divided in anticipation of Watershed Project approval and construction.

Interestingly, existing households were joining new nuclear families in claiming this land. Although the rationale for this was not clear it is possible that in some instances new land for *khang chung* (minor house) development was being claimed by the existing *khang pa* (main house), thus safeguarding ownership should the *khang chung* at a later date be disbanded. It is also possible that the participation of existing main households in the division of new land implies a disproportionate cornering of project benefits.

Equity, or rather a lack of, is a major WDP policy issue and it is feasible that existing wealthy (and thus powerful) households are competing with poorer, newly established and potentially land- less *khang chungs* for new agricultural plots.

Whilst the Watershed Guidelines stress the involvement of all beneficiaries in the selection, planning and execution of the Project, this inclusion was not reflected by feedback from the village-level Watershed Associations. Members of WAs were frequently uncertain about the final action plan, the project budget, the release of funds, project duration, and the procedure for asset maintenance. In some instances informants claimed that their respective Watershed Committees - which usually constituted the educated and entrepreneurial village elite - were 'voted' in through less than democratic means. The motivation for this may simply be that several senior members of the WC are entitled to a monthly salary. However, villagers did imply that the assimilation of the WDP into existing relations of power and status may have longer-term implications for the cornering of benefits by a village minority.

The potential for a competitive environment at the village-level is represented by point y. Admittedly it is too early to comment conclusively on these perceptions, however, Shah does note the ability of WDPs to show "a clear hierarchy of benefits and beneficiaries". In Zangskar the lack of knowledge and general awareness at the village level of the Watershed ideology, and even basic Guidelines, certainly impedes the WA in its role of overseeing the actions of the WC. Indeed, such an asymmetric distribution of knowledge also has fundamental implications for the sustainability of these development projects. Inequalities were also reported with reference to the employment of beneficiaries in project construction.

A prosperous village in Central Zangskar consisting of over 70 households has already completed one of their watershed objectives, yet not all households were in a position to offer labour. Payment for labour - which was made in a cash lump sum - was, perhaps understandably, restricted to contributing households. Whereas a traditionally organised village work party accounts for labour-short households, employment on the Watershed Project made no such provision. Consequently, the disadvantaged houses that were unable to contribute labour simply missed out.

The resulting division and ill-feeling was amplified by a disproportionately large payment for what amounted to 5 days of "easy work". The cornering of Watershed development benefits, often by those who need them the least, is also observed by Ninan, (1988: 30-38) who, with reference to WDPs in Uttar Pradesh. and Karnataka, discusses similar employment trends which exclude marginal and disadvantaged groups.

Patterns of inequality resulting from WDPs have also been recognised by entire settlements in Zangskar as the principal, larger settlements have the potential to corner more resources than their contiguous hamlets. For example the WDP in Shillingskit - a hamlet of sTongde - falls under the control of sTongde's WC. The geographical separation of these two settlements has already resulted in children from Shillingskit being unable to benefit from SECPAD's entry point activity of offering winter tuition.

Shillingskit's inhabitants however were more disappointed than angry, implying that it was better for an inequitable idea to benefit a minority rather than a situation where everyone lost out. Shah attributes this approach to a history of one-off development initiatives and superficial relationships between PIAs and the village-level. It would be rash to make assumptions on just one example but the following section also lends weight to the idea that some Zangskaris are accepting projects and their benefits simply because they are being given away, and not because of the project's merit.

The inheritance of project assets and the accompanying responsibility of maintenance is crucial to the sustainability of WDPs. The frequently cited mechanism for encouraging this is the participation of beneficiaries in both the planning and construction of their own projects. Admittedly, Zangskar's WDPs are still in their infancy, however, at this early stage there would appear to be a widespread absence of awareness and understanding as to the Watershed ideals of asset ownership and the obligation of maintenance.

The trends initiated by BDO and PWD subsidies look set to continue as the majority of Zangskari farmers stated that once irrigation projects were completed they would still require government materials and funds to undertake maintenance work. A PIA also confirmed this approach by stating that "it is beyond the capability of the village to maintain projects; villagers are poor and now that there are wages people will not work on or maintain projects without getting paid".

The BDO was considered to be the best option for assistance with Watershed Project maintenance although several farmers stated that if this was not forthcoming then the community would resign itself to undertaking the work. Although traditionally organized village maintenance still exists in Zangskar, the new cash economy, combined with a history of 'give-away' subsidies, would appear to have begun the erosion of this system. A curious paradox arises where Zangskaris profess that irrigation is fundamental to their self-reliance, yet, at the same time they would appear to be increasingly dependent on aid to construct and maintain their irrigation channels. One Zangskari farmer did however offer a possible explanation by expressing concern over this dependency yet at the same time praising the benefits of the Watershed's employment opportunities.

Kerr *et al* (1996: 11) highlights the ability of WDP subsidies to promote eager participation in wage labour regardless of the project nature, an observation which was supported by several farmers in Lungnak who perceived employment to be more important than their irrigation needs. This approach was graphically illustrated in one village where, regardless of completion, a number of farmers - who were also WC members - were already planning to abandon their irrigation project should it go over budget.

The situation encountered in Kumi, arguably Zangskar's most water scarce village, offered a further insight into the ready acceptance of subsidised projects. Despite having already committed labour and resources to moving the village - and thus utilising the water security offered by the Lungnak River-Kumi's Watershed Project is addressing the old problem of linking receding snowfields with the existing village.

Farmers were divided over the best course of action; some confessed that the future of Kumi lies in moving to the new village site and others claim the most important objective is improving the failing system above the existing village. The PIA (ZDPS) was apparently reluctant to engage in channel work at the new site so villagers also approached the PWD and the BDO.

As a result Kumi now has two projects; the proposed WDP pipe/channel linking the snowfields with the existing village and the 3km PWD/BDO canal to the new village site. The justification for two apparently conflicting aims was that the Watershed Project would offer a transition period, temporarily providing Kumi with water security whilst the new settlement and adjoining fields were being completed. Having received little beneficial attention over the last 20 water-scarce years, the inhabitants of Kumi have made numerous unaided and unsuccessful attempts at improving the efficiency of their irrigation channels. In the last five years a significant effort has also been made to develop and irrigate the new village site. Considering the chronic nature of their water scarcity it is perhaps understandable that Kumi households are now accepting any assistance that is offered.

The majority of households from water scarce villages such as Kumi and Pishu have already contributed a significant share of labour into WDP projects. Pishu's 25 households have invested an average of

25 days labour this year, although some individuals claimed to have been working on their Watershed Project for over 2 months. The ability of farmers in Central Zangskar to commit such a large block of time was found to be very rare. The majority of settlements visited (where WDPs were in operation) were only prepared to contribute several days of labour in the spring and then a slightly longer period between the harvest and the first winter snow. When discussing employment on Watershed Projects, Zangskari farmers often highlighted the constraints imposed by a shortage of both labour and time.

The brief summer provides the only opportunity to perform a host of activities which prepare the Zangskaris for the long months of winter isolation. Thus, when communities were visited in June and July, work on their respective Watershed Projects was far from being a priority. Furthermore, smaller settlements clearly struggle to offer labour at any time of the year.

In Shillingskit, which only has 7 inhabited houses, not even the lure of employment would free up time from agricultural duties and house maintenance. In response, SECPAD, Shillingskit's PIA, were organising a group of Nepali labourers to construct the village's proposed canal. The occurrence of labour short villages which are unable to have a direct stake in their Project assets again has implications for the future lack of project maintenance.

PIAs and the Private Sector

As registered NGOs, all three of Zangskar's PIAs have differing degrees of experience and areas of expertise. Both SECPAD and LYA were established in 1994 to administer overseas funds which were supporting two private schools in Central Zangskar.

Since their conception they have also branched respectively into rural initiatives such as the production of handicrafts and village-level social welfare. ZDPS has seven years experience in mainly youth activities such as school tournaments and a winter ski school. All NGOs admit that despite their recent selection as PIAs, their involvement in Watershed activities is a new challenge in areas where they have little previous experience.

Given that all of Zangskar's Watershed Projects have another three or four years to run it would be premature to pass judgement on the PIAs ability to oversee successful WDPs. However, since their selection as PIAs in October 2001 (LYA was selected at the later date of November 2002) their village activities would appear not to be entirely compatible with the Watershed ideology.

PIAs were united in their belief that the most positive aspect of the WDP was its "direct contact with the Zangskari people". The Programme was diplomatically described as being more transparent and more equitable than the existing system of development subsidies distributed through Government Organisations (GOs). The re-empowerment of Zangskaris to plan and work on their own projects, especially as contact with Kargil is limited to the summer months, was also perceived to be a positive step towards Zangskar's increased autonomy.

The scope to develop irrigation efficiency was viewed as a priority, not only because of recent water scarcity but also because irrigation is seen as the corner stone of Zangskar's sustainable development, especially in terms of the creation of a rural economy through cash crops.

The PIAs also linked irrigation to the benefits of afforestation and the creation of local willow plantations, a convenient and non-labour demanding source of fuel, fodder and construction materials. The apparent lack of attention aimed at social development may indicate a common feature in WDPs where funds are diverted into high cost structures at the expense of a more holistic, but less tangible approach. Indeed, Zangskari perceptions of how their water security can be improved focus on channel construction and material assets (productive efficiency) rather than the invisible economic solutions of allocative efficiency.

Whilst PIA objectives conform with the 2001 Watershed Guidelines they do not appear to directly equate to activities observed on the ground. At the village-level there was a mixed response to the implementation activities of the PIAs. Several villagers were very positive stating that PIAs were operating in the best interests of the village. The majority however were less enthusiastic with complaints ranging from inactivity and the withholding of work funds to a general lack of expertise and village contact.

A common observation was that the urgency of village needs was not mirrored by PIA activity. Villagers highlighted that the senior members of NGOs/PIAs receive Watershed funds as well as government pensions and incomes from contracting businesses and other "unknown" sources. Rather than benefiting the selected villages the WDP was hinted at being just one more business opportunity for PIAs, a view that was particularly prevalent in one village which was not involved in the WDP.

This emerging picture was put into perspective by the observations of an NGO employee in the Indus Valley who observed that some of Ladakh's NGOs are not built on ethical or reflexive ideals. In Ladakh, employment with an NGO is often perceived as the next best thing to the salary and status of Government employment. In Zangskar the NGOs would certainly appear to attract entrepreneurial, ex-government employees and those that have fallen short of making Government service.

Although village-level observations do not fully classify the PIAs as a private sector organisation, position z mapped onto the group/grid diagram recognises what would best be described as the PIAs private sector potential. It would be unfounded to imply that PIAs were solely involved in WDPs to make a profit. However, their activities on the ground do imply that there is more to their motivation than simply improving village livelihoods.

Villagers recounted what amounted to competition between two PIAs over one water scarce village and perhaps more importantly the village's potential to repeatedly attract development funding. Informants also commented on the general lack of village consultation (there was no mention of PRA activity), training and the occasional use of unqualified personnel to make engineering decisions. In several instances the salaried Watershed Development Team had yet to materialise in its entirety and villagers were frustrated at nearly two years of inactivity.

There were also reported discrepancies over project design and a reluctance to use expensive materials, although considering the poor success of existing concrete and cement channels, PIAs where perhaps well founded in prescribing, where appropriate, the use of traditional (low cost and low technology) earth channels. Despite the shortcomings of the PIAs it is also important to recognise that operating at the interface between Zangskar's Buddhist villages and Kargil District's Muslim administration is inherently problematic. The slow release of funds from the DDA has undoubtedly hampered PIA activity and may in part be responsible for accusations that funds are being misappropriated.

Indeed a similar incident is reported by Sheikh (2002: 33) in the Indus Valley where PIAs were taking the blame for the DDA's withholding of funds. Sheikh, who heads the Leh-based NGO/PIA Ladakh Development Organisation, also recognises the lack of provision made by the DDA for the training and capacity building of Ladakh's PIAs.

These sentiments were echoed by a member of a PIA in Zangskar who claimed that training from the DDA would improve their efforts, especially as the Watershed Guidelines made little allowance for Zangskar isolation and climatic constraints.

It is possible to see the predominantly Buddhist Zangskar Sub-Division as being a troublesome thorn in the side of Kargil's Muslim administration. Placating Zangskar's demands and minimising the political cost of what amounts to little more than administration - is undoubtedly a resource intensive practice. Yet, in this volatile border region the maintenance of internal stability is highly desirable, especially considering the recent history of unrest and agitation, largely resulting from issues of poverty, the politicisation of religion and political neglect. The government (both Central and State) reaction has been to contain the Ladakh-wide situation through a combination of measures including the granting of AHDC, encouragement of tourism, and the introduction of a range of development initiatives and subsidies.

Due to the centrality of irrigation, and possibly the immediate benefits that water harvesting structures usually yield, irrigation development has been a prominent Government initiative across Ladakh. In Zangskar this is not only reflected by the focus of the PWD and BDO but also by the nature of approved Watershed Projects. Successful irrigation projects (of which there are few in Zangskar) are politically beneficial icons of government concern and activity.

Indeed, Gyawali & Dixit (1999: 67) note the political legitimacy that irrigation development yields when recipients, such as those in Zangskar, have been exposed to the democratic ideal of the welfare state. Add to this the re-empowerment and funding of communities to sustainedly manage their own resources and development, and the introduction of the WDP to Zangskar has significant political potential.

Indeed, upon further assessment it is possible to recognise several further political benefits associated with WDP implementation. Considering the activity of the PWD, BDO, the Agricultural Department, and the staple foods and kerosene available at the government store, Zangskar has undoubtedly cornered a share of available subsidies. However, the fact that everything is transported into Zangskar by road places a large financial burden on Kargil's administration and the development funds filtering down from State and Central Government.

The potentially sustainable WDPs now being introduced to Zangskar may therefore be viewed as offering Kargil's administration an opportunity of easing this burden of financial responsibility. Without providing an alternative, the political cost of phasing out subsidies, which in some of Zangskar's villages have almost become institutionalised, would be severe.

In the long term however, by promoting independence and

sustainable socio-economic development, the WDP decreases, at least in theory, the reliance of remote Zangskari communities on external resources.

Although an initial investment from State and Central Government is required to cover the project implementation, there is also the potential for a degree of cost recovery as the long-term dependency on subsidy should ultimately decrease. Thus, the arrival of the WDP can be viewed, by State and District administration, as creating a window of opportunity which allows a shift in development policy and the steady reduction of Kargil's financial obligation to Zangskar.

Whilst the WDP has the potential in the long term to relieve Kargil of its financial obligation it immediately eases the administrative responsibility of managing Zangskar's development.

To an extent Zangskar's PIAs now act as a buffer between village-level Zangskaris and the Kargil administration. PIAs head Zangskar's Watershed Programmes and although many Zangskaris are aware that the funding source and Project approval still rests with the Government, the PIAs are now the focus of attention (and blame) which was once directed at Block and District level administration. In fact the DRDA-administered DDA is now in a position to criticise PIAs and the management of the resources which were once under its direct supervision.

It is important to note however that whilst the responsibilities of local resource management and funding allocation have been effectively decentralised, the villages and the PIAs continue to be under the control of Kargil District and J & K State. Despite the political benefits of the WDP, the transfer of responsibility to PIAs and the village-level has not been smooth. Sheikh accuses the DDA of struggling to make the transition between 'implementor' and 'facilitator'.

Another member of an NGO (PIA) in Leh also implied that at the District-level there was a 'general bureaucratic displeasure' at having to relinquish the power associated with decision-making. At the time of fieldwork it was unclear what the relationship was between Zangskar's PIAs and Kargil's administration but the erratic and slow release of funding would perhaps imply that the transferral of funds to the PIAs and the village-level was not a priority. In Zangskar, current irrigation development through the WDP is between the 'having' and 'operating' phases of innovation and change. Whilst Zangskaris *have* the Programme in place, its effective operation *has* yet to be achieved.

The sustainable Watershed ideology is intended to equitably improve socio-economic conditions through the participation of beneficiaries.

However, rather than facilitating sustainable irrigation development, the WDP in Zangskar is being modified and reshaped by the incompatible perceptions, and the relatively short-term agendas, of the parties involved in implementation.

Excited by subsidies entering directly at the village-level, and the perception that the Programme is compatible with perpetuating Zangskari self-reliance and independence, the WDP is being readily embraced by Zangskar's farmers and PIAs.

The respective projects have the potential to offer valuable village-level employment opportunities and tangibly offset the immediate problems of drought and population increase.

Yet, at the same time, the subsidized nature of the Programme appears to be undermining the potential sustainability of resulting irrigation projects.

At this early stage the established dependence on existing subsidy would appear to be inhibiting the reintroduction of notions surrounding community ownership and asset maintenance. The Zangskari focus on improving village water security exists, yet it is frequently obscured by the short-term desire for employment, the blind acceptance of give-away subsidy and the shortage of available time and labour.

In the light of failing irrigation projects, this conflicting situation may in part be *due to the reluctance, and in some instances inability, of farmers in Zangskar to trade the immediacy of short-term benefits for the uncertainty of long-term returns.*

From a governmental perspective the implementation of the WDP in Zangskar is a politically useful tool. However, at the village and PIA level, an absence of Watershed Programme understanding, combined with the easing of the District's responsibility to administer development, has created a situation where there is little accountability.

The asymmetric dissemination of the Watershed Guidelines undoubtedly limits the participation and monitoring ability of Watershed Associations.

The resulting lack of transparency has the potential to fuel rumours and accusations at both the PIA and village-level. The subsequent loss of trust between implementing parties may prove to be especially problematic as the WDP in Zangskar has created and legitimised a social and political space through which personal profit and political ambition may be pursued.

At this early stage it is not clear to what extent this space is being utilised. However, inequitable patterns of employment, and the possible cornering of benefits and decision making by an empowered entrepreneurial minority, would suggest that the Watershed Programme is perceived by some as being more than an opportunity to improve village water security.

Admittedly, Zangskar's WDPs are in their infancy and the shortcomings outlined in this study may be remedied with experience. Indeed, an increase in transparency and accountability between the parties involved may well be all that is required to install equity and sustainability back into Zangskar's Watershed Programmes.

Chapter 3

New Approach to India's Rural Drought Problems

REINVENTING WATERSHEDS PROPOSING

India's substantial investment in rural development during the past five decades of the post-independence era has not produced a corresponding transformation on the ground. With the focus of development being largely sector specific, the impact has rarely reached the rural poor. Consequently, sustainable transformation of rural ecosystems with people's participation has remained isolated and scattered. People's participation, the key to community-centered development, has remained largely stuck in the 'you participate in my programme' mode. As a result, for lack of community ownership the impact of investment in rural development and natural resource generation has not been effective and long lasting.

The recent impetus to support already-existing watershed development programmes is a policy response to reverse the prevailing malaise in integrated rural development programmes. The experiences, as of now, are quite varied, ranging from about a dozen very successful projects to a large number of non-effective ones. But wherever enterprising community groups had decided to better their circumstances with some external assistance, they were able to improve their socio-economic conditions considerably in a relatively short time span. Watershed development programmes in India and worldwide are increasingly emphasising flexibility, participatory processes and institution building so as to fulfill "all the major demands of the Agenda 21" in one shot. While this is a major step towards sustainable development, people's participation alone may not yield the desired results unless the process adequately addresses the key issues.

LARGE SCALE COVERAGE BUT POOR IMPACT

The Watershed Development Project (WDP), funded by India's

Ministry of Rural Areas and Employment, has been implemented since 1994–95 in about 10,000 watersheds in all the major states of the country.

In addition to an estimated US $ 3.5 billion being invested for rehabilitation and development of micro-watersheds by the government, the World Bank, Danida, Sida, SDC, DFID and the Indo-German Watershed Development Programme are supporting WDPs through direct funding support as well.

Watershed programmes are increasing being favoured by most of India's state governments. Of these projects, launched during the past six years, Andhra Pradesh state in the south of India accounts for 24 per cent with Madhya Pradesh, the central Indian state, a close second with a 17 per cent share of these projects.

Uttar Pradesh (10 per cent), Gujarat (8.6 per cent) and Tamil Nadu (7 per cent) are the other states who figure prominently in the list. Ironically, several districts of both Andhra Pradesh and Madhya Pradesh are currently reeling under drought. However, it is not the coverage that is significant but the impact and the long-term sustainability of the projects.

Despite the best of donor support for soil and water conservation based on the watershed approach, land degradation continues unabated. A study by the Centre for Economic and Social Studies indicates that the rate of degradation of land in rainfed areas in the 1990s is likely to have proceeded at more than twice the rate observed in the 1980s, basically on account of uninterrupted soil erosion.

Furthermore, the continuing shortage of drinking water in several states of the country clearly demonstrates that the existing interventions for drought proofing have failed to neutralise the overall processes of land degradation in rural areas. Despite repeated failures, the thrust on WDPs has yet to fade way. Failure to evaluate the strategies and approaches adopted by these projects has in fact led to proliferation of WDPs in the country.

With donor support clearly in its favour, most implementing agencies have found an easy escape route to hide failures. Lack of cross-referencing between donors has helped the implementing agencies continue to access resources despite failures of their previous projects.

Unless governments and donors wake up to this reality; appraise and evaluate current strategies; and set into motion approaches that will overcome the shortcomings, three decades of watershed development programmes will all be wasted away. Not counting on its technicalities, the watershed approach can be used as a political tool to eradicate poverty in rural areas.

GUIDELINES FAVOUR FIXED TREATMENT RECIPE

The entire watershed development programme is based on guidelines that typically reflect the per hectare cost of land treatment. Though these guidelines get reviewed periodically, the review is often to accommodate cost escalations and to set revised targets. Fixed costs often disregard the wide variability in biophysical and socio-economic conditions. Consequently, most projects do not adapt to the local conditions. Fixed guidelines do not allow cross-learning among the wealth of approaches already existing. Given the nature of current watershed projects, investment is channeled into diverse activities (both on-farm and off-farm) that often involve trade-off across resources as well as individual stakeholders.

The increase in the number of activities is often to account for the previous lapses that exclude some of the biotic and/or abiotic resources from the intended benefits of the project. Consequently, the resultant impact is not only slow but at times not even tangible. With times and resources getting distributed thinly across several activities, it requires fine tuning between physical and social engineering for desired results. A typical project may include several components under the following key areas of intervention: water resources development, socio-economic development, crop planning & agriculture, household energy, forestry issues and livestock & animal husbandry.

It is now clear that the watershed approach has gone much beyond the scientifically determined methods of soil and water conservation. In doing so, not only has per hectare cost of conservation gone up but also a range of diverse activities has come under its fold. Given this holistic approach, watershed development programmes have become more complex to implement than some of the individual component-based schemes. To give the existing WDPs a new focus, some donors have come up with 'watershed plus' – an improvisation over the existing watershed approach to accommodate livelihood concerns.

A recent study of a watershed project in Chhattisgarh state revealed the predicament of the implementing agency to complete diverse activities within the stipulated time frame. Subsidies were uniformly spread, irrespective of a household's economic status; and the distribution of benefits was determined largely by the land-base of the households rather than being based on negotiations between households. Such skewed distribution of benefits has impacted negatively on the project's ownership and sustainability.

In contrast, a project like Sukhomajri (a village in the Shiwalik hills in north-west India in the state of Haryana, where an innovative watershed project was completed in the early 1980's) used water as a community resource to help design a system that did not follow the inequity in land holdings and yet ensured water security for all.

With rights over impounded water in the three check dams equally shared by both landed and the landless in the village, the benefits of rainwater harvesting were equally shared between the community. Furthermore, the project ensured that a portion of the incremental gain is ploughed back to create a social capital. Social capital, in the case of Sukhomajri, holds the key to the project's long-term sustenance. Incidentally, the concept of creation of social capital has rarely been translated into the new generation watershed projects. Unlike most projects, the landless in Sukhomajri village harvested the benefits of the project by selling their share of water to the needy.

A sound land care system, based on the principle of social fencing, was not only capable of regenerating biotic resources but triggered a range of farm and non-farm activities as an outcome, which were not part of the original project design. It was then for the community to make informed choices for utilising the rejuvenated natural resources to their benefit. The 'watershed plus' approach of incorporating a range of activities, on the other hand, puts pressure on resources and manpower for their timely execution.

Though the diverse activities may add value to the project, it is often at the cost of innovation. Community watershed programmes currently under implementation demonstrate the need to slim the number of activities to adapt to local biophysical and socio-economic conditions and bring in innovative approaches that ensure reviving the community to nurture and utilise its natural resources on a sustainable basis.

INEQUITABLE SHARING LEADS TO POOR SUSTENANCE

With the success rate of current technology-heavy watershed projects no more than 25 per cent, there is a need to re-visit and re-assess the entire strategy of the watershed approach towards drought proofing and food security. A study in Gujarat found that checkdams, the most favoured technology for watershed projects, only benefited 15-16 per cent of the households directly. Whilst the benefits of checkdams are more or less sure, benefits to individual farmers from other structures such as stream plugs or contour bunds may not be so immediate and/or substantial. Yet, a significant portion of the entire project cost is invested in creating checkdams.

Though it depends on size and the catchment area, a majority of checkdams account for an average of 50 per cent of the entire project cost. The remaining funds get thinly distributed over other components of the project. Though projects are often abuzz with host of social activities including self-help groups, income-generating activities etc, the benefits get shared between few families only. Can a community that hasn't benefited be expected to contribute towards sustenance of the project? The baggage of activities in a watershed project has become too heavy. Reduction in the number of activities in favour of those that provide maximum benefit is being considered one option to bring down the per hectare cost of land treatment.

Activities should get selected on the basis of their expense and impact in the project area. Since the community contribution (approximately 10 per cent) to the total project cost is a mandatory requirement by a majority of donors, choice of most-benefiting activities will give legitimacy to their (community) contribution as they will know the return on their investment. Once a value is attached to each activity, the community can be expected to plough back a portion of the incremental gain to create social capital.

Chakriya Vikas Pranali in Jharkhand state is one live example that is based on sharing of benefits in a continuous 1:3:3:3 sharing system: 10 per cent of the income goes to the village welfare fund meant for the villagers in dire straits; 30 per cent goes to the owner of the land from which the income is derived; 30 per cent to the tiller; and 30 per cent to the village fund for other socio-economic activities in the village.

Not only the reinvestment from the rejuvenated land resources helps sustain the system, it empowers the community to help spread and replicate the concept to nearby villages. In most other projects, the one-time investment is used to accomplish the project activities to the best of its abilities. Such projects rarely go beyond the 'contractor' approach of completing tasks, oblivious of what might happen to the project later.

Building watershed harvesting structures as part of the watershed development projects is a very easy task – any contractor endowed with a bit of money can do it. But building an effective structure, which starts off a process of self-management and self-regeneration in a village community, is a much more difficult task.

Watershed projects can gain a lot by paying attention to the revival of traditional water harvesting structures. Not only is this less costly, community's contribution and participation is easily accomplished in such initiatives. Furthermore, reviving community structures leads to rebirth of community spirit and community management – an aspect on which substantial time and resources are spent in watershed projects. If the continuing drought-like situation in western region of the country is any indication, India's three decades of efforts to drought-proof the country through the watershed approach haven't really worked. Failure of investments notwithstanding, substantial budgetary provisions have been made by the central and state governments for the rehabilitation and development of micro-watersheds.

The increased frequency of droughts has necessitated a fresh thinking on the watershed approach to drought proofing. There are limited options though. Doubts have already been expressed on the basic flaws in the design and implementation of watershed projects. Despite frequent review of the guidelines to incorporate compelling concerns, watershed projects fail to deliver the promised impact. Since the benefits of watershed projects do not get shared amongst a large number of households, its adoption by the community continues to remain a major concern. The technology-heavy nature of these projects on the other hand distances the community even further.

In contrast, projects that have done better in the recent past were based on community's traditional water harvesting and conservation practices. Not only did these projects ensure better community participation, the cost of implementation was on the lower side, too. As these people-centred programmes benefit large sections of the society, its operation, management and sharing of benefits are based on equity and ecological principles. In contrast, the WDPs show a clear hierarchy of benefits and beneficiaries, with households receiving irrigation benefiting most, followed by those getting on-farm treatments like field bunds, and the landless with or without livestock, receiving least.

Since this hierarchical pattern is largely governed by the given geo-hydrological profile, superimposed on the existing structure of land ownership and associated water rights, the observed inequality in the distribution of project benefits is fairly obvious. What is particularly concerning is that these issues are treated as more or less structurally determined, rather than being placed at the centre of a participatory process. The need is to initiate negotiations between different beneficiaries and stakeholders.

ASSESSING AGRICULTURAL SUSTAINABILITY IN WATERSHEDS IN INDIA USING GEOMATICS

Rain-fed agro-ecological regions (AER) which encompass the semi-arid tropics (SAT) and hot dry and moist sub-humid regions of India, extends over 76.74 M ha out of the total geog. area of 90.4 M ha in the states of Andhra Pradesh (AP), Maharashtra (MAHA), Karnataka (KAR) and Tamilnadu (TN) in Peninsular India. Watershed-based development has been an important component in the schemes for economic planning for development for ensuring agricultural productivity, natural resources management and enhancing rural livelihood. Over 71 million persons depend on agriculture and allied activities in these states alone according to 2001 Census of India (GoI,2001).

Annual rainfall in this vast region ranges from 400 mm in the arid belt of Rayalseema in southern Andhra Pradesh to over 800 mm in the rest of the region which occur in 52 - 55 rainy days. Over 50 % of this rainfall which occurs in summer is associated with thunderstorm activity which lasts for a few hours. Considering such a rainfall pattern, rain - water harvesting, storing and prudent use for agriculture and other allied activities is of utmost importance.

Watershed Development Programme was initiated in 1980s to address these limitations of the rain-fed agro-climatic and ecological niches of India. The emphasis and approach of the projects vary in consonance the ecological nuances of the regions. In scanty rainfall regions like in case of Maharastra uplands where annual rainfall is 500 mm, increasing irrigation potential by water harvesting and recharge of groundwater is given precedence.

In case of Telengana region where annual rainfall is 800 mm and the terrain is gently undulating, water harvesting and surface storage in soil profile and farm ponds in preferred for enhancing productivity of rain-fed crops. In case of arid regions of Rayalseema where rainfall is scanty, water harvesting in field tanks for irrigation during crop stress, are preferred modes. The soil and water conservation structures (S&WC) viz., check-dam, stone weirs, contour bund, live bunds, vegetative cover, key-line plantation, grass way etc. are planned to provide impediments to overland runoff which induce soil erosion and deplete nutrients from agricultural fields.

The structures are laid to guide runoff to designated farm ponds and tanks for water harvesting on surface besides impounding water for facilitating deep percolation for groundwater recharge. Also, soil degradation is a matter of grave concern in India which threatens agriculture. In Peninsular India over 47.3 M ha area have poor degraded soils of which 0.15 M ha have been categorized as wasteland.

More recently, the frequency of abnormal trends in climate like late onset of monsoon, early withdrawal, long inter-seasonal breaks, shift in peak rainfall period, cloud-burst, thunder storms, etc., are increasing leading to drought or flood occurrences. Thus, watershed development programme (WSP) is undeniably, the most comprehensive programme for achieving agricultural and ecological sustainability in the rain-fed regions of India.

WATERSHED DEVELOPMENT PROGRAMME (WDP) IN INDIA

A primary reason in favour of watershed-based development (WSD) in rain-fed region is the enormous cost of major water projects like the under-construction Narmada river-valley project. Hence the Govt. of India lays emphasis on augmenting water resources through small and decentralized projects like the watersheds.

In India watershed projects have been undertaken under six major national programs, viz. Drought-Prone Area Programme (DPAP), Desert Development Programme (DDP), National Watershed Development Project for Rainfed Area (NWDPRA), Watershed Development in Shifting Cultivation Areas (WDSCA), Integrated Watershed Development Project (IWDP) and Employment Assurance Scheme (EAS) etc. by three Central Ministries of Govt. of India namely, Ministry of Rural Development (MoRD), Agriculture (MoA) and Environment & Forestry (MoEF). Significantly, 70% of funds for watershed development in the country are being spent under these six major programs. In 1999 a common approach was evolved for undertaking of WDP in India.

Several evaluation studies have been undertaken to assess the impact of WDP in rain-fed regions of India. Notable among them are by John Kerr & N. K. Sanghi (1992), John Kerr et al., (2002), Hanumantha Rao (2000) and Ratna Reddy et al. (2004). Most of these evaluation studies conducted till date, are based on econometric analysis as baseline information is often very difficult to obtain from the project implementing agency or the funding agency without which meaningful evaluation is impossible.

The present study attempts to use tools of Geomatics in consonance with conventional techniques of field survey, soil and water analysis and socio-economic survey for creating baseline information of pre-project period and evaluation of post-project scenario.

STUDY AREA

The methodology was applied to evaluate watersheds developed in the Telengana region of AP which forms a part of the semi-arid and dry-sub-humid agro-ecological region of India (Velayutham et al, 1999). Five watersheds located in Telengana region in the districts of Rangareddy (RR) and Nalgonda were studied. The present chapter describes the evaluation of WSD in one of the villages, Pamana in Chevella Mandal in RR district. While over 5000 ha area around Chevella township was developed as model watershed in 1985, Pamana village/ watershed (7807'30"E & 17016'45"N) was developed within Chevella Mandal under DPAP programme in 1999. Under the WDP, ten check- dams were constructed in two micro-watersheds (MWS, catchment of first-order stream) in a planned manner.

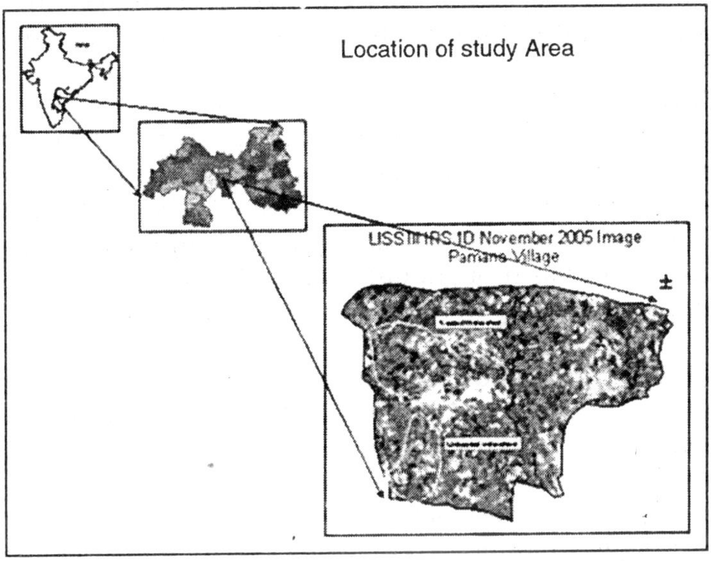

The study of a treated and another untreated micro-watershed in Pamana provided interesting results. The village has a total area of 1010 ha with 300 farm households. Land holding size range from 0.5 - 4 ha although a majority of land holding size is lesser than 1 ha. A large majority of them are un-irrigated. Soil is essentially black belonging to Vertisols.

Earlier, paddy cultivation was dominant wherever irrigation was available, however, with initiation of WDP, cash crops viz., cotton, tomato, carrot, coriander, onion, sugarcane, garlic and maize, besides flowers like Marigold, Jasmine, Chrysanthemums and Aster, have been introduced. In accordance with new guidelines for implementation of WSD in participatory mode, a Watershed Committee was formed in the village consisting of stakeholders (farmers belonging to the respective watersheds).

Seed money was sanctioned to the village administration (Gram Panchayat) and the Department of Agriculture (DoA, AP Govt.) guided the villagers to identify and locate check-dams and other S&WC structures. The physical structures were built by the villagers under the guidance of engineers from the State Department of Govt. of AP, and payments were made from the DPAP fund.

This was a two-pronged strategy under which the impact of drought was to be minimized by supporting farmers by providing livelihood during drought and ensuring agricultural productivity through S&WC on the long-run.

The entire village has 19 micro-watersheds (MWS) of which 4 are treated with S&WC structures. For the present study we have taken one treated (TMWS) and another untreated micro-watershed (UTMWS).

The TMWS selected for this study has an area of 129.49 ha. Over 40% of its area is irrigated, 51% un-irrigated where rainfed agriculture is practiced and the rest is scrubland with the ownership of Govt. The UTMW selected for the study covers an area of 21 ha with 23 % under irrigated agriculture and the rest under rainfed cultivation. The study of these two MWS has brought out interesting results.

MATERIAL & METHODS

Study was conducted in an integrated modular mode (Fig. 2) using Geomatics tools - GIS, GPS and Satellite data in addition to conventional techniques used in multi-disciplinary studies viz., field survey using toposheets, transact walk, civil survey using Total Station and GPS, soil survey, sampling and physico-chemical analysis in the lab, socio-economic survey using questionnaires and open-ended interviews with farmers, Participatory Rural Appraisal (PRA) at the village and collection of secondary data from mandal and district head quarters etc.

Watershed hierarchy was identified and mapped using ArcGIS (Ver. 9.0). The catchments of 1st order stream were nested into that of 2nd order stream catchments which in turn were nested into the 3rd order stream catchments and so forth.

The National Watershed Atlas classifies the entire drainage network of India in a hierarchical manner consisting of 5 stages, namely, water resource region, river basin, catchments, sub-catchments and watershed; we have delineated the 3rd, 2nd and 1st order streams and up-linked these to the national watershed hierarchy in earlier studies.

Watersheds in Pamana village were delineated in a similar fashion into sub-watersheds, mini- and micro-watersheds respectively. A treated or developed micro-watersheds (TMWS) where S&WC structures were located and one untreated micro-watersheds (UTMWS) where no interventions were made, were selected for evaluation. The nomenclature of the selected treated micro-watershed is as follows: TMW – 4D1E8a4c which stands for 4 – Southern Water Region, D - Krishna River, 1 – Musa River, E – Himayatsagar sub-catchment, 8 – Watershed No. (4th order stream catchments), a sub-watershed code (3rd order stream catchments), 4 mini-watershed (2nd order stream catchments) and c micro-watershed (1st order stream catchments).

Socio-economic survey was conducted using a structured questionnaire that was created to measure sustainability indicators. Two databases – one for watershed information and another for village information - were created in MS-Access and were linked to GIS platform.

Soil and water sampling and analysis for 12 physico-chemical and biological parameters were undertaken and the results were input into Watershed Database. Satellite imagery of IRS 1D LISS-III of 16 Nov. 1998 (pre-project period), 17 March 2001, 16 Oct. 2004 and IRS – P6 data of 4 Nov. 2005 (post-project period) were digitally interpreted using ERDAS Imagine Ver. 9.0 and land use, cover and degradation maps were prepared. The results were integrated for evaluation of WDP in Pamana.

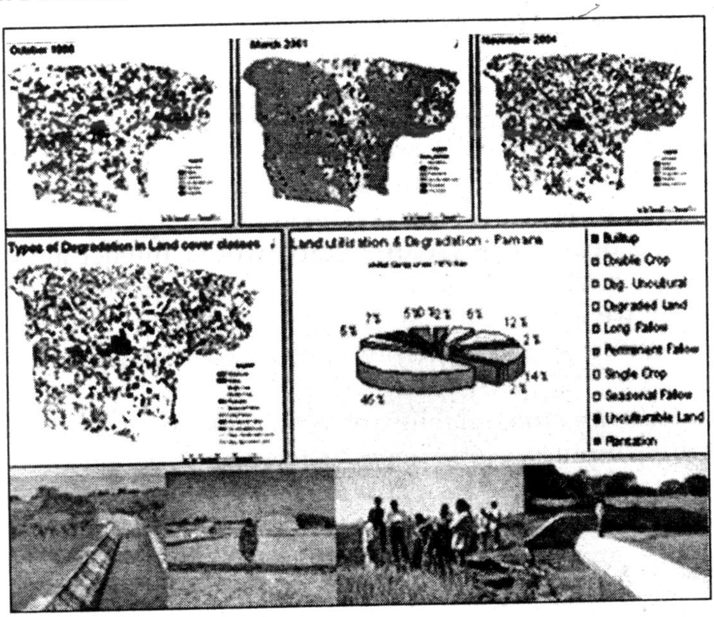

METHODOLOGY FOR EVALUATION OF WDP IN PAMANA

For evaluation of sustainability of WDP in Pamana village over 120 soils and water samples were collected and analysed for assessing soil fertility status and water quality. Over 180 farm households were interviewed and a socio-economic analysis was carried out using structured questionnaires to generate baseline information for testing the relevant sustainability indicator. Forty eight sustainability indicators were assessed for 8 issues at 5 levels of analysis as indicated below.

Table. List of Sustainability Indicators used for Evaluation of WDP

S. No.	Issues of Agricultural Sustainability	Sustainability Indicators	Leve/Spatial Extent of analysis				
			Household	Farm land	MWS	Village	AER
1	Rescurce availability	6	3	3	5	6	3
2	Intensity & diversity of land use	4	1	3	4	4	-
3	Land quality	11	2	10	10	7	-
4	Land practices by farmers	9	-	2	2	2	-
5	Land users awareness & institutional capacity	9	3	6	7	9	-
6	Soceital value of farm land	2	2	2	2	2	-
7	Extent of risk burrering	12	4	3	8	8	1
8	Equity in society 2	2	2	2 .	2	2	2
	Total	48	17	31	40	40	6

Suitable indicators were selected for assessing the five tenets of sustainability. For instance, to assess sustainability of agricultural productivity, crop yield was used as an indicator. To evaluate security of livelihood operations, factors which would impact agricultural production systems viz., soil cover, yield variability and rainfall distribution, were to be analysed. For environmental protection, soil quality/ quantity, water quality/ quantity, biodiversity, etc., were to be evaluated.

In order to ascertain viability of agricultural operations, net farm profitability, input-use efficiency, off-farm income and economic return were to be studied. To ascertain whether WDP and its components of development strategy were acceptable, types of conservation practices prevalent and adopted by farmers, method of Land Use Planning and decision – making criteria of farmers, were to be analysed.

Significantly, the methodology enabled the quantification of several aspects associated with land management in developed / treated watersheds, viz., new land management practices introduced and adopted, changes in cropping pattern and replacement of traditional cropping system with cash crops, availability of credit facility from Rural Co-operative Banks or credit society and support from institutions like Agricultural Extension Department etc., which could all ensure agricultural sustainability.

A Principal Component Analysis (PCA) was attempted on the set of sustainability indicators selected for the study. Sustainability indicators pertaining to three aspects i.e., Farmers' Satisfaction, Resource Conservation and Watershed Management were used for assessing agricultural sustainability in the selected watershed under the present study.

Data generated under various modules of the project were correlated with the list of sustainability indicators chosen for the purpose. A matrix was evolved and averages, threshold values and indices for each of the indicator were generated adapting the method indicated by Gomez et al., (1996). For instance, under the issue of resource availability where deforestation rate could be numerically estimated using satellite data of various time-period, availability of fuel wood or fodder could be assessed only qualitatively.

Hence scores were generated to depict the status of sustainability. A final scorecard was generated to indicate how various components of WDP had faired in the MWS in Pamana; which components were strong, which were the weak links that could make the WDP unsustainable. Cob - web diagrams used by Gomez et al., (1996) were found appropriate to draw attention to the strength and weaknesses of the WDP undertaken in Pamana.

RESULTS & DISCUSSION

The issue of sustainability of rainfed agriculture was analysed using a wide range of indicators which included a few indicators that can be deduced from satellite data and analysed using GIS. While the viability of WDP was assessed using indicators pertaining to Farmer's Satisfaction, Resource Conservation and activities initiated under Watershed Development programme, as indicated through cob-web diagrams; use of satellite data and GIS for study of aspects of LUCC (fig. 6) and NDVI have been discussed here. The issue of equity was also analysed for the TMWS & UTMWS in the study area. It was found that equity was better addressed in TMWS (0.36) which is slightly poorer than the national average for India (0.30 - 0.34) while in UTMWS (0.47) the inequality among farmers was severe.

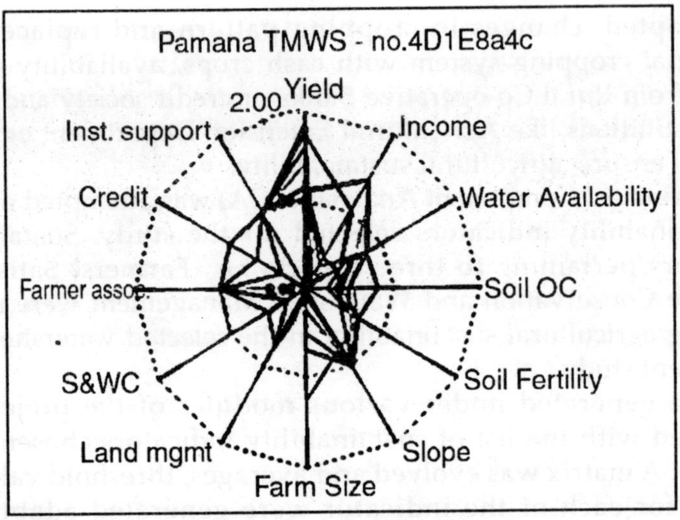

Fig. Cob-Web Diagram

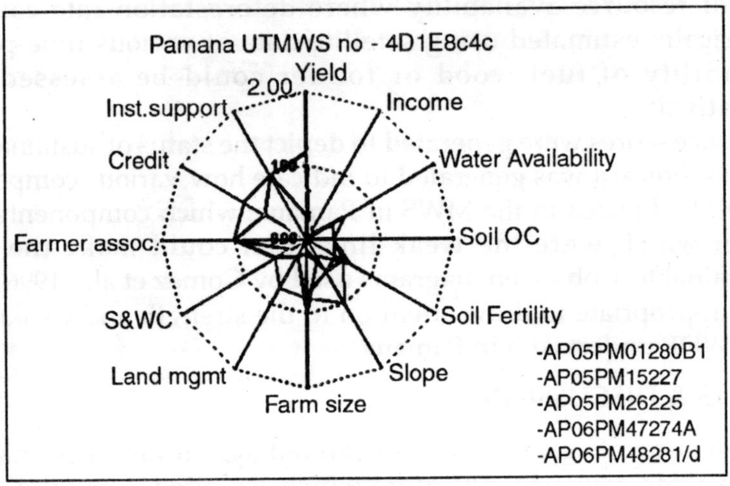

Fig. Cob-Web Diagram

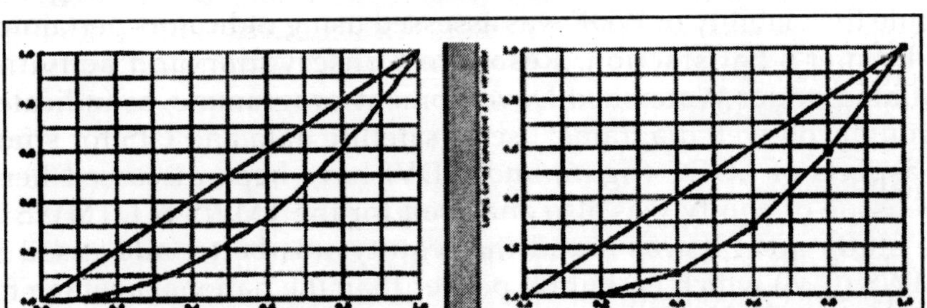

Fig. Equity Ananlysis using Lorenz Curve for Finding Gini Coefficient

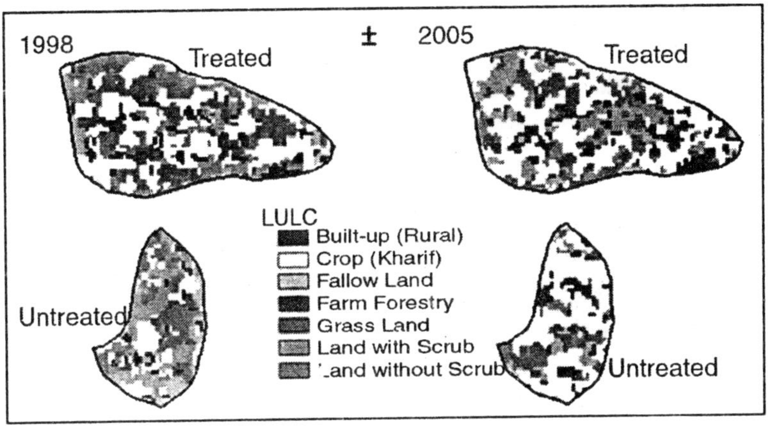

Fig. Land use and Types of Cover in TMWS & UTMWS in pre-& post WD Period in Pamana

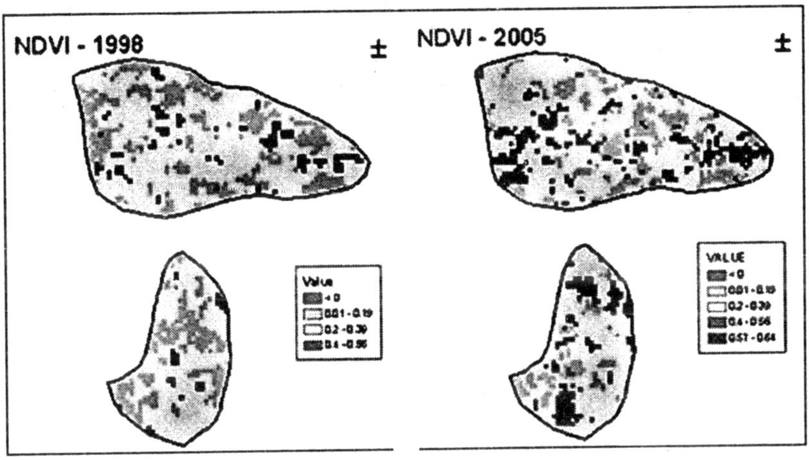

Fig. NDVI in pre-& post WD period in Pamana

In order to study the impact of sustainability of WDP satellite imagery of the study area of pre-WS period and Post-WS period were classified using ERDAS Imagine Ver. 9.0. While agricultural area is largely stabilized in the TMWS there is a large expansion of agricultural activity in marginal lands belonging to LCC IV and V with slope of 4 - 5 % that would induce soil erosion.

Land fallow which is essential for rejuvenation of land is being maintained in the TMWS, while there is no fallowing practiced in UTMWS. Tree groves have been depleted in UTMWS and vegetative cover from scrublands have also decreased which would adversely affect soil OC content. All these trends indicate unsustainable trend in UTMWS.

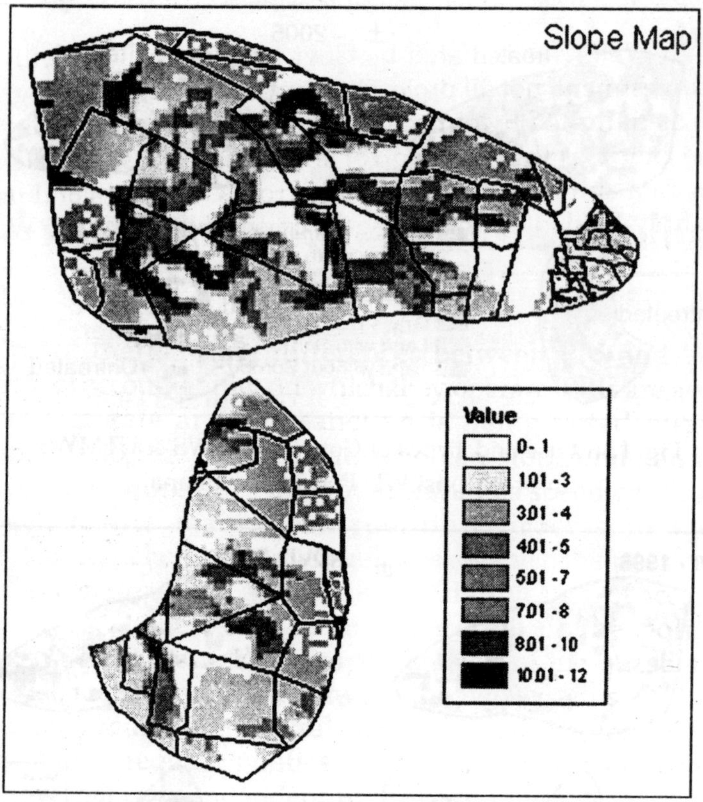

Fig.Slope in TMWS & UTMWS in Pamana

Normalized Difference Vegetation Index (NDVI)

Light reflectance from vegetation was compared in the post-kharif season period in 1998 and 2005. NDVI was calculated using the formulae IR–IR/IR+IR where IR stands for infrared and R stands for red portion of the EM spectrum.

The Vegetation Index (VI) was normalized for topographic effect and the difference of NDVI in the study area in pre- and post-watershed project period is indicated in figure.

Evidently in pre-watershed period, NDVI in both TMWS and UTMWS was lower. In 2005, NDVI was higher in large parts of TMWS which indicates better rabi-crop stand. Higher NDVI was also recorded in UTMWS in Nov. 2005 which is indicative of good vegetation stand.

WDP is essential for development of rain fed agriculture in India as it would ensure agricultural production through natural resource conservation viz., water harvesting and halting of soil erosion.

Although WDP was initiated by Govt. of India in 1985 with massive public investment, not all projects are yielding sustainable agricultural growth as indicated by our methodology.

The merit of this method for evaluation lies in the fact that it derives data and information heavily from Geomatics and uses them successfully to identify the strengths and weakness of the WDP in an area.

The methodology permits temporal analysis of indicators, comparison of various WDP across AER and at different periods of time. Most importantly, it helps identify weak linkages in a given WDP which can be corrected to achieve agricultural sustainability in rain fed areas.

Chapter 4

Participatory Watershed Development in India

The last decade has seen the increasing decentralization of responsibilities for management of natural resources to the community level. The WSD programmes being supported by the Government of India are one example of this. Microwatershed development is currently attracting over £300M/yr of central government funding alone, with a further £100M – £200M/yr from the States and from donor-supported projects. Numerous projects are also being implemented through NGOs. Approaches to assessing the success of WSD in India have evolved over time.

During the 1970s and early 80s, the evaluation of watershed initiatives was based primarily on biophysical criteria. Large amounts of money for instance were invested in monitoring soil losses, changes in vegetative cover and changes in depth of water tables. The late 1980s saw a growing awareness that WSD is about more than maintaining or improving the productivity of natural resources. The guidelines issued by the then Ministry of Rural Areas and Employment for example cover multiple objectives including:

- Productive (optimum utilisation of the watershed's natural resources).
- Social (employment generation and development of other economic resources in the village).
- Ecological/environmental (easy and affordable solutions that build on indigenous knowledge).
- Equity (emphasis on improving the economic and social condition of the resource poor).

Current concerns centre on the issue of sustainability. WSD is fundamentally about the creation of new opportunities for all sections of the community, which should be sustainable beyond the rehabilitation period. This chapter uses the sustainable livelihoods approach to take a fresh look at the impact of WSD on rural livelihoods.

It focuses particularly on questions relating to the extent to which WSD activities result in the creation of new livelihood opportunities and the extent to which these opportunities are both equitably distributed and sustainable.

The chapter begins by introducing the policy and institutional context for WSD in India. The chapter concludes with an assessment of how successful WSD has been as an instrument for reducing rural poverty, and provides some suggestions on how the impact of WSD on livelihoods might be enhanced.

THE POLICY AND INSTITUTIONAL CONTEXT

This section provides an overview of the evolution of WSD programmes in India and the roles that organisations and policies have played. It describes the evolution in thinking on WSD that has moved away from a focus on the rehabilitation of the natural resource base towards a more holistic vision which sees WSD as one approach to tackling rural poverty.

Government – Early Initiatives

Largely in response to the substantial benefits recorded from 'special projects' of the kind described in Box, the GoI initiated a number of efforts attempting to implement similar approaches on a wide scale. These included principally the National Watershed Development Project for Rainfed Areas (NWDPRA) under the Ministry of Agriculture and Cooperation, and under the Ministry of Rural Areas and Employment, the Drought Prone Areas Project, the Desert Development Project, the Integrated Wastelands Development Project (IWDP), and part of the Employment Assurance Guarantee Scheme.

Box: Benefits generated by special watershed projects in the 1970s and 1980s The case of Kharaiya Nala, southern Uttar Pradesh

Close collaboration between a GoI research organisation and State line departments (Agriculture, Horticulture, Animal Production) and the village community allowed the testing and implementation of what has now become a 'standard' package of WSD treatments: ridge-to-valley approach, including contour trenches on the upper slopes; fencing of common land to allow regeneration; checkdams and gullyplugging, and the promotion of soil and water conservation on agricultural land.

The benefits recorded 3 years after completion of the activities included:

- Reduction in runoff from 70% to 22%, and in soil loss from 41t/ha to 1.9t/ha.

- Rise in water table from 12m to 6m depth; increase in number of wells dug by villagers from 12 to 483; increase in proportion of irrigated land from 9.6% to 69%.
- Area sown under kharif and rabi increased by 85% and 233%; cultivation in zaid for first time; extensive switch to higher value crops.
- Over 300 ha of arable wasteland brought into crop production;
- Seven-fold increase in fuelwood offtake; switch from cowdung to firewood as fuel.
- Fodder planted on almost 150 ha of field bunds and gully plugs; dry fodder production increased by 235%.
- Switch from livestock species previously relying on open grazing (nondescript cattle; sheep; goats) towards higher quality stall-fed draught and milch animals.
- Former herders find employment as agricultural labourers;
- Increased soil fertility.
- Creation of village investment funds from the sale of produce from the commons (fuel, fodder) and creation of a revolving fund for small loans at nominal interest.

These early large-scale approaches to WSD exhibited a number of flaws:

- They were primarily evaluated according to biophysical factors. Thus success was related?to changes in yield, land use, livestock productivity etc. There was no attempt to understand what changes this brought about in people's livelihoods.
- The approach adopted in different projects WSD projects largely reflected the primary mandate of?the responsible Ministry.
- They operated in a top-down fashion, implementing 'models' developed through research, with?little attempt to take into account the needs of local users of watershed resources, nor to adapt to local biophysical or socio-economic conditions.
- Practically no contribution from beneficiaries was required, even on private land, so that there was?little local ownership of the new works, little commitment to their maintenance and a syndrome of dependency on government was strengthened.

As a consequence, their performance was poor, and a major survey by Kerr et al. (1998) in semi- arid areas, came to the conclusions that:

- Recently completed watershed projects funded under central government schemes and including projects supported by the

World Bank, showed no noticeable difference in most criteria from untreated villages and plots.

- Soil conservation works on rainfed plots are much better maintained where farmers themselves?have made a financial contribution towards their cost:
  - Participatory approaches combined with careful selection of villages with favourable characteristics for WSD offer the best prospects of success. On a small scale they attract special treatment from NGOs and/or donors which may be difficult to sustain as they attempt to expand, but may in part be compensated by innovative NGO-GO collaboration.
  - Women and the poor receive some benefit from employment created during rehabilitation, but?otherwise their interests are widely neglected. Continued vigilance is needed to protect their interests.

Even prior to Kerr's study, growing concerns in some government circles had led to the drafting of the new Guidelines on Watershed management, which aimed to strengthen participatory planning and local-level decision-taking. These drew on the experience of a number of innovative NGO and donor-supported experiences, which are now summarized.

NGO APPROACHES TO WATERSHED DEVELOPMENT

It is generally accepted that NGOs have had a more successful record in WSD. NGOs are widely diverse in terms of size, professional qualifications, geographical and sectoral areas of work, and (in relation to WSD) in approaches, scales of financial support, networks and linkages. Drawing on the work of two of the more experienced NGOs in WSD – Mascarenhas (1998) for OUTREACH and Fernandez (1993) for MYRADA – the following account describes an approach characteristic of many NGOs.

NGOs typically attempt initially to support the formation of 'focus groups' of various kinds on the principle that groups which are moderately homogeneous in economic and social terms allow skills to be developed for the much more difficult negotiations over the management of CPRs in a WSD context. Such groups may centre on economic activities such as milk production, poultry rearing or the collection and processing of various non-timber forest products (NTFPs).

In many cases, they will not be related at all to natural resource management, for instance in cases of groups formed for "woman and child health" activities. The general intention is that groups should develop skills in leadership, negotiation and conflict resolution, and should, in a Freirean mode, develop a capability to identify the problems and opportunities they face within the wider social and economic context, respond to them from their own resources as far as possible, and learn how to draw in external resources as necessary.

Groups are typically also encouraged to develop savings and small-scale local lending schemes to assist in breaking the control of local money-lenders, and are given training in the basic skills necessary to manage funds. The major difficulty with NGO approaches to date is that they require lengthy periods of face-to- face interaction.

NGOs can allocate the resources for this since they simply mandate themselves to work in limited areas. By contrast, government supported programmes necessarily aim for wider coverage. Much of the current policy debate over WSD concerns ways of blending the depth of NGO initiatives with the breadth of government approaches.

Donor Agencies' Initiatives

In the Indian context, the importance of donor-supported efforts lies less in their overall financial contribution than in their flexibility to experiment with and 'pilot' new approaches – a flexibility which government norms and procedures do not easily permit. Donors have also placed a strong emphasis on developing partnerships between stakeholders in the WSD process.

The approach of the Swiss Agency for Development and Cooperation (SDC), for instance, is characterized by efforts to share responsibilities between NGOs and the public sector according to their comparative advantage, in ways which are mutually reinforcing. They also seek to achieve collaboration among government departments and search for coherence and complementarily between participatory appraisal methods and 'scientific' measurement and planning techniques used by government.

Closer NGO-government collaboration is potentially synergistic. It has featured in the rhetoric of GoI planning and, has succeeded in some sectors such as health. However, historically it has achieved limited progress in agricultural and natural resource development in India. SDC-supported projects such as, initially PIDOW in Karnataka and, subsequently, PAWDI in Rajasthan, have been unable to break down vested interests on both sides.

The Government Response

The Government of India's awareness of experimentation by both

NGOs and donors led to a number of innovative responses, principally by the MoRAE. New Guidelines for Watershed Development were published in 1994, intended to be 'common' across MoRAE. The Guidelines have found wide acceptance, and are innovative in:

- The extent to which they promote the principles of participatory WSD.
- The extent of decentralised decision-taking and funding they permit.
- Their provision for different types of organisation (government; NGO; university; even private?commercial) to act as local project implementing agencies.

They embody approaches to WSD intended to be:

- Implemented on a wide scale,?in ways which:
  - Enhance the productivity of the resources.
  - Ensure a degree of equity in the distribution of benefits.
  - And are ecologically and institutionally sustainable.

This section has documented the evolution of WSD approaches. There is no doubt that the most recent approaches recognise the importance of building on peoples' priorities and it is this factor which determines whether WSD can be judged successful. The following sections assess how, for whom and to what extent WSD is meeting the livelihood priorities of the rural poor.

WATERSHED DEVELOPMENT AND ITS IMPACT ON LIVELIHOODS

So what has been the impact of these new initiatives on the livelihoods of the poor? Given their relatively recent history, it is difficult to make a detailed assessment of the performance of WSD programmes. Anecdotal evidence and impact assessments of individual projects (particularly of donor and NGO-implemented projects) suggest several positive trends:

- Increase in cropping intensity and yields of both irrigated and dryland crops.
- Reduction in the threat of drought to crop and livestock production.
- Increase in milk production (livestock systems move from open grazing and towards cross-breeds).
- Recharge of ground water.
- Decline in sedimentation downstream.
- Improved fodder production.
- More livestock managed under stall-fed conditions.
- Year-round availability of drinking water.

- Creation of employment opportunities for landless labour.
- Diversification of the village economy into artisanal and other activities as people gain the?confidence to approach banks for credit.

This kind of evidence however tells us little about the actual impact of WSD on peoples' livelihoods. The key point here is that WSD is not an end in itself, but it is a means to an end: 'reducing the incidence of rural poverty'. The SL approach provides a framework for analysing the 'fit' between WSD activities, rural livelihoods and ultimately poverty reduction.

WATERSHED DEVELOPMENT AND CAPITAL ASSETS

This section assesses the extent to which WSD has enabled rural communities to build up their capital assets and the extent to which these gains have been equitably distributed. It pays particular attention to CPRs, which have long been an important focus of WSD and are an important asset for the poor.

BUILDING UP THE ASSET BASE

As WSD approaches have evolved from primarily externally imposed biophysical interventions towards more participatory approaches encompassing a broader range of activities so the potential impact of WSD on household assets has increased. WSD has implications for all five types of assets defined in the SL framework, for example:

- Human capital – through capacity building activities; participation in new institutions and processes.
- Social capital – through the formation of watershed committees, user groups and new or strengthened institutions.
- Financial capital – through the establishment of credit groups, the establishment of a WSD fund.
- Natural capital – through increases in trees, livestock, irrigated area, more productive land.
- Physical capital – through increase in irrigation facilities, soil and water conservation structures.

There is a growing awareness of the links between different capital assets. Investments in physical capital such as bunds, check dams and the re-vegetation of common lands for instance are relatively easy to achieve. The returns to physical investments of this type however will rapidly decline if appropriate investments in social and human capital are not also made to develop sustainable and equitable institutions to manage these assets. Similarly the idea of sequencing is important.

Some NGOs argue strongly that the local institutions which determine access to natural capital (e.g. common land) need to be regularised before WSD activities are undertaken.

ACCESS TO CAPITAL ASSETS

Although WSD can no doubt have quite significant positive impacts on the natural resource base, there are growing concerns over the distribution of these benefits. The principle of the ridge to valley approach for instance may work against the interests of the poor who often rely on the commons in the upper slopes much more heavily than do the better-off. To begin with water control structures in the lower slopes would privilege those (i.e. the better-off farmers) having reliable access to agricultural land.

The notion that the poor, who are denied access to the commons during the rehabilitation period, can benefit from casual, unskilled employment opportunities created by construction requirements in the lower slopes is appealing, but requires close monitoring; and there are important questions surrounding the sustainability of any of these gains.

Box given below highlights some of the other tensions, which exist between better-off and poor households, regarding access to both existing and newly created assets.

Box: Watershed development – capital assets and the poor

- WSD envisages the construction of a wide range of physical assets principally for soil and water management, such as bunds, checkdams, gullyplugs, ponds, shallow wells, and so on. However, there remain questions over access and sustainability for the longer term, especially if joint responsibility for maintenance is not clearly agreed. The better-off, being landholders, generally benefit disproportionately from an increase in groundwater levels brought about by soil and water conservation. Information on how any agricultural intensification impacts on employment opportunities for the poor remains sketchy.
- Common pool resources (CPRs) such as grazing or forest land constitute an important form of?natural capital and joint action is generally required to manage these. Spending pressure under government programmes means, however, that the formation of resource user associations is hasty, and that the views of the poor and women are rarely taken adequately into account. Long standing problems, such as encroachment remain unsolved. After rehabilitation, the

increased potential offtake of timber, fuel and fodder often attracts the attention of the better-off to CPRs, which, in degraded form, had generally been the preserve of the poor.

- Social and human capital is being strengthened through the formation and capacity building of focus?groups, watershed associations etc.. However, the short period allowed for preparatory activities, together with the lack of social organisational skills among (especially government) project implementing agencies (PIAs), means that careful approaches to human and social capital development are rarely implemented on a larger scale.
- In terms of financial capital, most forms of rural investments in India suffer from a 'handout' syndrome, in which rural people have long been regarded as the recipients of government munificence, with the consequence that they distinguish clearly between 'government projects' for which the government has planning, implementation and (in their view) maintenance responsibility from activities they undertake privately.

Common Pool Resources

CPRs represent a form of natural and social capital that individuals and communities can draw on in pursuit of their livelihood strategies. A study by Jodha (1986) concluded that CPRs make a key contribution to rural livelihoods and are critical for sustainable agricultural production in semi-arid areas. They form a part of rural peoples' strategies for adjusting to the harsh and stressful environment.

More recent evidence also supports this view. Surveys conducted for the DFID Karnataka Watershed Development Project show that three quarters of the rural poor in the watershed areas depend on CPRs for fuel, fodder and some food. The role of CPRs is particularly important in lean times, acting as a buffer in times of drought or other crises. During droughts, the collection of tubers and wild foods from the forest is an important coping mechanism. CPRs also provide independent sources of income to women, through, for instance, goat rearing, broom making, leaf plate making and collection of tendu leaves and tamarind. All of these occupations depend upon CPRs, either for fodder or raw materials.

WSD in India has its roots in the widespread concern over land degradation – with approximately 170 million hectares classified as degraded. Large areas of this land are classified as CPRs. 'Before and after' pictorial comparisons of WSD efforts often compare areas of barren rocky common land with productive pastures and shrub cover.

The assumption is generally that management regimes have deteriorated into open access in a 'tragedy of the commons' scenario. WSD projects have invested considerable efforts in establishing both the rules for access to such areas and in creating collaborative agreements for community management of such CPRs.

The key question is the extent to which the poor retain access to CPRs after WSD efforts have taken place. To take one example, a crucial element of many watershed projects has been restrictions on the use of common grazing areas during rehabilitation and thereafter to permit sustainable off-take.

Adolph and Turton (1998) note that such controls in a watershed project in Andhra Pradesh had different impacts on households of different socio-economic status. Landowners were able to compensate for the loss of grazing, through the substitution of crop residues, the availability of which had increased due to improved irrigation facilities. Landless livestock owners on the other hand were forced to sell their livestock raised similar concerns.

Lokur-Pangare (1998) reports more positively from another project in Maharashtra. Here the existing users of specific common land and water resources – not the whole community – made the decision regarding how they were to be managed. Landless livestock owners formed a 'resource user group' and had a primary say in decisions over grazing land management. Overall the key question seems to be whether the short-term losses in terms of access to CPRs are outweighed by the longer-term gains.

A second dimension of change often brought about by WSD projects is the privatization both de facto and de jure – of CPRs. This can work both in favour of the poor and against. Commonly, the wealthier bribe local officials to turn a blind eye to encroachment, whilst in other cases encroachment is the desperate action of poor and landless households. Encroachments are evident throughout rain-fed areas and are legalized by State governments from time to time. Integral to some WSD programmes is the granting of individual plots of previously common land (usually the most degraded land) to the landless and poor.

Many of the private land plantations developed under the Sida-funded PAHAL project are located on encroached land previously under the jurisdiction of the Revenue Department. The project recognised that a fair amount of consensus prevails about rights over these areas and proceeded with the plantations, despite the lack of formal legal entitlement.

WATERSHED DEVELOPMENT AND ITS IMPACT ON LIVELIHOOD STRATEGIES

This section examines the impact of watershed policies and structures on the livelihood strategies adopted by rural communities – particularly the poor. It highlights changes in the composition of Scoones' (1998) 'livelihood strategy portfolio' via: agricultural intensification/extensification, livelihood diversification and migration.

AGRICULTURAL INTENSIFICATION/EXTENSIFICATION

WSD can open up new opportunities for households to both intensify and extensify production, for example through:

- Increasing the frequency of cultivation increased groundwater recharge often permits cultivation of a second crop, or 'life-saving' irrigation of a main crop.
- Adopting new technologies, such as improved varieties of staple crops, higher levels of input use and higher yields; shifts to the cultivation of higher value crops in some areas.
- Expansion of cultivated areas – soil and water conservation and irrigation structures can bring new land into production.

All of the above tend to generate increased demands for labour. These may be provided initially from farm family sources and subsequently by hired labour. Beyond pilot projects. however it is not known, for instance, how far intensification/extensification generates employment opportunities for households – often poor and the landless – who depend on labouring for their livelihoods. Some of the most striking evidence of intensification in rainfed areas in India comes from the livestock sector.

Restrictions on access to CPRs – often promoted through WSD – have encouraged the move towards stall-feeding systems for both large and small ruminants. In the shorter term – when restrictions are most strict - households may become more dependent on crop residues. Some studies have also highlighted the importance of private resources, with increased numbers of trees on farmland and the cultivation of range grasses and legumes in niches such as field boundaries, private wastelands, terraces and bunds.

In the longer term as access restrictions are eased and their productivity improves, CPRs can become an important fodder resource for cut and carry systems. Overall this suggests a diminishing role for CPRs in grazing systems, but an increasing role as a source of cut fodder. As the previous section outlined not all households will however be able to intensify successfully – the poor may be unable to cope with the loss of access to CPRs in the shorter term.

There are also important intra-household dimensions to intensification strategies. While increases in agricultural productivity might occur as a result of WSD, men usually appropriate on-farm gains and the increased drudgery is disproportionately borne by women. With an increase in land productivity, men frequently choose to increase the production of cash crops such as sugarcane and cotton. It is important therefore that women are included in decisions over land use and crop planning, to ensure that household food requirements (vegetables, fruit and a variety of grains) are provided for adequately.

Box: Women are farmers too An important objective of WSD is increased land productivity. It is therefore 'farmers' who first come forward to participate in project activities. It is often the same farmers who are nominated to the local decision-making bodies – including the watershed committee – set up for the project. Women are nominated to these committees under a kind of 'reservation policy', implying that they are not members of the farming community in their own right. However, women are also farmers.

They are the primary food producers and contribute more hours of work and perform more tasks than men in agricultural production. They not only cultivate land belonging to their families and work as agricultural labourers, they also independently lease plots for cultivation. Women farmers have independent views about farming practices and can contribute significantly to the improvement of agriculture. Women reported that the shift to cash cropping resulting from increased water availability in rehabilitated watersheds had generated increased risk for their families.

Increased investment financed through credit, has resulted in a highly vulnerable economic condition. Given a choice, women would opt for enhanced stability and the cultivation of food grains. Since women rarely own land, they are not recognised by policy makers as farmers; the term 'farmer' is usually treated as masculine.

It seems that WSD can support strategies of intensification if careful analysis of the livelihood strategies of different households and individuals precedes decisions on project activities. But in countries such as India, rigid socio-cultural relations at village level may work against the poor. For poor groups (usually the low-caste and/or women) to assert their rights over resources as these become more productive through rehabilitation requires careful initial development of skills and confidence, robust and transparent procedures for decision-taking, monitoring of any departures from what has been agreed, and awareness among all concerned that, if necessary, resort to an external agency continues to be feasible. These conditions are difficult to put in place on a wide scale. Serious policy attention therefore also needs to be given to alternative strategies of diversification and migration.

Diversification

In the past decade recognition has grown that agricultural production is only one of the strategies that contribute to livelihoods. Rural households – particularly the poor – engage in a wide range of activities. In India there are important traditions, often underlain by formal and informal institutions, which dictate the types of activities that households and individuals may pursue. For instance, villagers' participation in many non-farm activities such as carpentry or basket making is dictated by strong caste-based institutions.

The collection of NTFPs has traditionally been an important livelihood strategy for tribal groups. WSD has supported these diversification strategies in a number of ways. NGO projects have paid specific attention to providing opportunities to the poor to diversify their livelihoods, often through the formation of self help groups for women, the landless and other marginal groups.

These groups undertake a number of activities ranging from traditional crafts (such as leaf plate making; weaving and basket making), mushroom cultivation, forestry activities and so on. However, as recent detailed preparations for a DFID Western Orissa Rural Livelihoods Project demonstrated, the prospects of such enterprises tend to be assessed on the basis of historical production and sales levels. Certainly, the evidence points to some prospect that local people could benefit if restrictions on their access to or sale of NTFPs were lifted but over the longer term, artisanal products generally face inelastic demand, so that the scope for any increased contribution to livelihoods is likely to be limited.

Migration

Migration is one of the most important means of diversifying rural livelihoods. Migration strategies in India vary widely: one or more family members may leave the resident household for varying periods of time and in so doing are able to make new and different contributions to its wellbeing. Migration may be seasonal or permanent; it may be point-to-point or circular in fashion; it may vary in distance of destination from within province, to within country or, occasionally, cross frontier. The contribution of migration to livelihoods will depend on various factors, including the seasonality of movement, the length of time spent away, assets and social structures and institutions allowing for women (if men migrate) and others to pursue activities previously reserved for men and household heads.

A starting point in assessing the role and implication of migration for livelihoods in any location is an understanding of the rationale behind migration. Broadly speaking we can observe a continuum between negative distress migration and positive livelihood migration in terms the capacity of migration to facilitate or constrain livelihood security.

In large areas of tribal India, migration occurs as a defensive coping strategy to cover existing debts and seasonal variability. Households have little choice if they want to survive. Migrants (often whole families) leave their villages after crop harvest and migrate to towns. Many travel long distances often to the same place each year. For instance, members of one village in Orissa migrate en masse to work in the construction industry in Dehra Dun, almost two thousand kilometres away. Migration is especially important for tribal groups and generally the livelihood strategies of poorer households with little or no land depend more heavily on migration.

Whilst it has long been accepted that migrant remittances constitute an important component of rural livelihoods, increasingly in many areas, earnings from migration form the primary source of cash income. Mosse et al. (1997) report from a study in tribal areas of western India that earnings from migration contribute on average 86% of cash income and that 80–90% of households have at least one migrating member.

Migration is often highly formalized and involves complex sets of relations between building contractors, village recruiters and workers. These involve arrangements for securing work and providing for travel and accommodation expenses. These relationships are often highly exploitative in nature. For instance in western Orissa, labour contractors are also villagers' only source of loans. Highly indebted to the contractors, migrants have little bargaining power and are trapped in a vicious circle.

Migration is likely to become an increasingly important livelihood strategy for the rural poor. In relation to WSD there are several key issues which need to be addressed:

- Migrants' absence from villages tends to marginalise them from the decision-making process.
- Migrants' dependence on CPRs is highly seasonal and management strategies should accommodate this.

- Migrants may be absent at key points in the year and may lose out in the sharing of benefits from common resources.
- Migration often leads to the absence of the most able-bodied working population – often males– this has key implication for women's roles and decision making.

Migration has an important impact on participation in local level decision making processes. WSD involves the establishment of new institutions such as user groups and watershed committees. Migrants – generally the poorest – are often absent from villages and so tend to be marginalised from decisions on resource use. In the context of WSD in India, most NGOs and several donors see seasonal migration as a negative phenomenon, largely on account of its exploitative nature and its disruptive effect on family life and wider social relations.

This view clearly has some validity, but, whilst policy interventions should be sensitive to these requirements, they should not be unduly influenced by unrealistic ideals. WSD offers an opportunity to address migration issues: to promote intensification under WSD in employment-creating rather than in labour saving modes, so that substantial employment is created beyond the construction phase of rehabilitation (and it should be noted that labourers are unlikely to switch out of migration strategies until they are confident that substantial blocks of employment – probably at least forty days – are available in the home area); the second is to support migration processes so that labourers' conditions are improved, and they are gradually released from exploitative contractual arrangements.

Groups and committees set up by WSD programmes could provide an entry point for improving awareness of migrants rights and building their capacity to exercise them. A livelihoods perspective can reveal why certain groups fall out or do not participate in WSD, and how participation can be enhanced. For example, activities may fit badly with livelihood strategies of the poor if involvement is particularly risky, requires access to specific assets, or occurs at a time when other livelihood activities are a priority for them.

Some barriers may be insuperable, but in other cases, participation can be encouraged by adapting procedures to suit the livelihoods of the poor. Livelihoods analysis for an Indian watersheds project highlighted many obstacles for the poor in attending community meetings, relating to migration, lack of time, and power. As a result, the second phase of the project is paying careful attention to timing and developing alternative mechanisms for participation, including informal sub-watershed associations, and 'livelihood support teams' to work with the most vulnerable.

A SL perspective provides an opportunity to stand back and explore how WSD affects the livelihoods of the poor, and to see how these impacts can be enhanced. It highlights how even small changes in timing, location or sequencing can enhance the fit of WSD with livelihood priorities. Adopting a SL perspective has also encouraged a more explicit analysis of the ways in which WSD directly and indirectly affects people's livelihoods.

In terms of the overall impact of WSD – as a means to rural poverty reduction – watershed-based approaches have been demonstrated to lead to substantial improvements in rural livelihoods. However, they are not a panacea: productivity gains in pilot projects have not been achieved to the same extent on a wide scale and the links between productivity gains and livelihoods is complex and poorly understood.

Of most concern, is that achieving these productivity gains can actually work against the livelihood strategies of certain groups. For instance restricting access to CPRs to improve grassland producitivity denies the poor access to a valuable resource. Shifts to cash cropping, possible as a result of improved access to irrigation, can encourage a more risky livelihood strategy and lead to increased food insecurity at the household level.

Undoubtedly, to achieve distributional equity is the most difficult challenge. An adequate distribution of benefits (poor/better-off; men/women) relies on exceptionally careful and continuing vigilance. The final section explores how these challenges may be addressed to improve the 'fit' of WSD with the livelihoods of the poor.

Improving the Impact of Watershed Development

The conditions necessary for successful WSD are becoming clearly-established, and include: adequately sloping topography; adequate rainfall; high contribution of land-based activities to livelihoods; the willingness of landowners to allow the landless adequate access to the benefits of rehabilitation; adequate prospects for long-term ecological sustainability, including restrictions on the distribution and volume of groundwater extraction; adequate prospects for long-term institutional sustainability, including market links, and strong community commitment to joint action in resource management.

Clearer guidance is needed for choosing between WSD and other options. There is also a need to improve the design, implementation and follow-up of WSD projects.

Opportunities for improving the design, implementation and follow-up fall into 3 broad categories:

- Strengthening human capital at two broad levels: first, a

longer pre-investment period to create?and support local 'focus' groups; second, stronger capacity building for the staff of government agencies and *panchayat raj* institutions.

- Strengthening financial capital: the creation of small, locally managed savings and credit?groups allows poor people to break free from moneylenders; more broadly, insistence on beneficiaries' contributions to rehabilitation, especially on private land – and accompanied by reforms to allow banks loans for this – will help to break the syndrome of dependence on government.

Finally, more thought is needed on development pathways appropriate for those areas where preconditions do not favour WSD. For example, seasonal migration is an important livelihood strategy for semi arid India. Efforts to make the process less exploitative and support it through the public sector, via interested NGOs and/or via support through the banking system are likely to generate substantial poverty impact. By contrast, the evidence on alternative economic opportunities which are not land-based is mixed.

NGO lore is replete with examples of how poor groups have succeeded in producing and marketing products made from NTFPs, such as tendu-leaf plates, or in launching new enterprises such as beekeeping or silk production. However, these rarely consider the costs of providing financial capital and the necessary skills, and of underwriting the high level of risk. Where these are taken into account, the prospects for micro-enterprise appear much weaker. In conclusion, SL analysis (drawing on a wide variety of methods) should be used to encourage a broader and more structured assessment of impacts relevant to the poor. This can help practitioners and policy-makers to adjust their approaches and so enhance poverty impacts.

DRAINAGE

Drainage is the natural or artificial removal of surface and sub-surface water from an area. Many agricultural soils need drainage to improve production or to manage water supplies. The earliest archaeological record of an advanced system of drainage comes from the Indus Valley Civilization from around 3100 BC in what is now Pakistan and North India. The ancient Indus systems of sewerage and drainage that were developed and used in cities throughout the civilization were far more advanced than any found in contemporary urban sites in the Middle East and even more efficient than those in some areas of modern Pakistan and India today. All houses in the major cities of Harappa and Mohenjo-daro had access to water and drainage facilities. Waste water was directed to covered drains, which lined the major streets.

DRAINAGE IN THE 19TH CENTURY

This operation is always best performed in spring or summer, when the ground is dry. Main drains ought to be made in every part of the field where a cross-cut or open drain was formerly wanted; they ought to be cut four feet (1.2 m) deep, upon an average. This completely secures them from the possibility of being damaged by the treading of horses or cattle, and being so far below the small drains, clears the water finely out of them. In every situation, pipe-turfs for the main drains, if they can be had, are preferable. If good stiff clay, a single row of pipe-turf; if sandy, a double row. When pipe-turf cannot be got conveniently, a good wedge drain may answer well, when the subsoil is a strong, stiff clay; but if the subsoil be only moderately so, a thorn drain, with couples below, will do still better; and if the subsoil is very sandy, except pipes can be had, it is in vain to attempt under-draining the field by any other method.

It may be necessary to mention here that the size of the main drains ought to be regulated according to the length and declivity of the run, and the quantity of water to be carried off by them. It is always safe, however, to have the main drains large, and plenty of them; for economy here seldom turns out well. Having finished the main drains, proceed next to make a small drain in every furrow of the field if the ridges formerly have not been less than fifteen feet (5 m) wide.

But if that should be the case, first level the ridges, and make the drains in the best direction, and at such a distance from each other as may be thought necessary. If the water rises well in the bottom of the drains, they ought to be cut three feet (1 m) deep, and in this ease would dry the field sufficiently well, although they were from twenty-five to thirty feet (8 to 10 m) asunder; but if the water does not draw well to the bottom of the drains, two feet (0.6 m) will be a sufficient deepness for the pipe-drain, and two and a half feet (1 m) for the wedge drain.

In no case ought they to be shallower where the field has been previously leveled. In this instance, however, as the surface water is carried off chiefly by the water sinking immediately into the top of the drains, it will be necessary to have the drains much nearer each other—say from fifteen to twenty feet (5 to 6 m). If the ridges are more than fifteen feet (5 m) wide, however broad and irregular they may be, follow invariably the line of the old furrows, as the best direction for the drains; and, where they are high-gathered ridges, from twenty to twenty-four inches will be a sufficient depth for the pipe-drain, and from twenty-four to thirty inches for the wedge-drain. Particular care should be taken in connecting the small and main drains together, so that the water may have a gentle declivity, with free access into the main drains.

When the drains are finished, the ridges are cleaved down upon the drains by the plough; and where they had been very high formerly, a second clearing may be given; but it is better not to level the ridges too much, for by allowing them to retain a little of their former shape, the ground being lowest immediately where the drains are, the surface water collects upon the top of the drains; and, by shrinking into them, gets freely away.

After the field is thus finished, run the new ridges across the small drains, making them about ten feet (3 m) broad, and continue afterwards to plough the field in the same manner as dry land. It is evident from the above method of draining that the expense will vary very much, according to the quantity of main drains necessary for the field, the distance of the small drains from each other, and the distance the turf is to be carried.

The advantage resulting from under-draining, is very great, for besides a considerable saving annually of water furrowing, cross cutting, etc., the land can often be ploughed and sown to advantage, both in the spring and in the fall of the year, when otherwise it would be found quite impracticable; every species of drilled crops, such as beans, potatoes, turnips, etc., can be cultivated successfully; and every species, both of green and white crops, is less apt to fail in wet and untoward seasons.

Wherever a burst of water appears in any particular spot, the sure and certain way of getting quit of such an evil is to dig hollow drains to such a depth below the surface as is required by the fall or level that can be gained, and by the quantity of water expected to proceed from the burst or spring.

Having ascertained the extent of water to be carried off, taken the necessary levels, and cleared a mouth or loading passage for the water, begin the drain at the extremity next to that leader, and go on with the work till the top of the spring is touched, which probably will accomplish the intended object. But if it should not be completely accomplished, run off from the main drain with such a number of branches as may be required to intercept the water, and in this way disappointment will hardly be experienced.

Drains, to be substantially useful, should seldom be less than three feet (1 m) in depth, twenty or twenty four inches thereof to be close packed with stones or wood, according to circumstances. The former are the best materials, but in many places are not to be got in sufficient quantities; recourse therefore, must often be made to the latter, though not so effectual or durable.

It is of vast importance to fill up drains as fast as they are dug out; because, if left open for any length of time, the earth is not only apt to fall in but the sides get into a broken, irregular state, which cannot afterwards be completely rectified. A proper covering of straw or sod should be put upon the top of the materials, to keep the surface earth from mixing with them; and where wood is the material used for filling up, a double degree of attention is necessary, otherwise the proposed improvement may be effectually frustrated.

The pit method of draining is a very effectual one, if executed with judgment. When it is sufficiently ascertained where the bed of water is deposited, which can easily be done by boring with an auger, sink a pit into the place of a size which will allow a man freely to work within its bounds. Dig this pit of such a depth as to reach the bed of the water meant to be carried off; and when this depth is attained, which is easily discerned by the rising of the water, fill up the pit with great land-stones and carry off the water by a stout drain to some adjoining ditch or mouth, whence it may proceed to the nearest river.

CURRENT PRACTICES

Modern drainage systems incorporate geotextile filters that retain and prevent fine grains of soil from passing into and clogging the drain. Geotextiles are synthetic textile fabrics specially manufactured for civil and environmental engineering applications. Geotextiles are designed to retain fine soil particles while allowing water to pass through. In a typical drainage system they would be laid along a trench which would then be filled with coarse granular material: gravel, sea shells, stone or rock.

The geotextile is then folded over the top of the stone and the trench is then covered by soil. Groundwater seeps through the geotextile and flows within the stone to an outfall. In high groundwater conditions a perforated plastic (PVC or PE) pipe is laid along the base of the drain to increase the volume of water transported in the drain.

Alternatively, prefabricated plastic drainage systems made of HDPE called SmartDitch, often incorporating geotextile, coco fibre or rag filters can be considered. The use of these materials has become increasingly more common due to their ease of use which eliminates the need for transporting and laying stone drainage aggregate which is invariably more expensive than a synthetic drain and concrete liners. Over the past 30 years geotextile and PVC filters have become the most commonly used soil filter media. They are cheap to produce and easy to lay, with factory controlled properties that ensure long term filtration performance even in fine silty soil conditions.

21st Century Alternatives

Seattle's Public Utilities created a pilot programme called Street Edge Alternatives (SEA Streets) Project. The project focuses on designing a system "to provide drainage that more closely mimics the natural landscape prior to development than traditional piped systems". The streets are characterized by ditches along the side of the roadway, with plantings designed throughout the area.

An emphasis on non curbed sidewalks allows water to flow more freely into the areas of permeable surface on the side of the streets. Because of the plantings the run off water from the urban area does not all directly go into the ground but can also be absorbed into the surrounding environment. According to the monitoring by Seattle Public Utilities, they report at 99 per cent reduction of storm water leaving the drainage project.

DRAINAGE IN CONSTRUCTION

The civil engineer or site engineer is responsible for drainage in construction projects. They set out from the plans all the roads, Street gutters, drainage, culverts and sewers involved in construction operations. During the construction of the work on site he/she will set out all the necessary levels for each of the previously mentioned factors. Site engineers work alongside architects and construction managers, supervisors, planners, quantity surveyors, the general workforce, as well as subcontractors. Typically, most jurisdictions have some body of drainage law to govern to what degree a landowner can alter the drainage from his parcel.

REASONS FOR ARTIFICIAL DRAINAGE

Wetland soils may need drainage to be used for agriculture. In the northern USA and Europe, glaciation created numerous small lakes which gradually filled with humus to make marshes. Some of these were drained using open ditches and trenches to make mucklands, which are primarily used for high value crops such as vegetables. The largest project of this type in the world has been in process for centuries in the Netherlands. The area between Amsterdam, Haarlem and Leiden was, in prehistoric times swampland and small lakes. Turf cutting (Peat mining), subsidence and shoreline erosion gradually caused the formation of one large lake, the Haarlemmermeer, or lake of Haarlem.

The invention of wind powered pumping engines in the 15th century permitted drainage of some of the marginal land, but the final drainage of the lake had to await the design of large, steam powered pumps and agreements between regional authorities.

The elimination of the lake occurred between 1849 and 1852, creating thousands of km^2 of new land. Coastal plains and river deltas may have seasonally or permanently high water tables and must have drainage improvements if they are to be used for agriculture. An example is the flatwoods citrus-growing region of Florida. After periods of high rainfall, drainage pumps are employed to prevent damage to the citrus groves from overly wet soils. Rice production requires complete control of water, as fields need to be flooded or drained at different stages of the crop cycle. The Netherlands has also led the way in this type of drainage, not only to drain lowland along the shore, but actually pushing back the sea until the original nation has been greatly enlarged.

In moist climates, soils may be adequate for cropping with the exception that they become waterlogged for brief periods each year, from snow melt or from heavy rains. Soils that are predominantly clay will pass water very slowly downward, meanwhile plant roots suffocate because the excessive water around the roots eliminates air movement through the soil.

Other soils may have an impervious layer of mineralized soil, called a hardpan or relatively impervious rock layers may underlie shallow soils. Drainage is especially important in tree fruit production. Soils that are otherwise excellent may be waterlogged for a week of the year, which is sufficient to kill fruit trees and cost the productivity of the land until replacements can be established. In each of these cases appropriate drainage carries off temporary flushes of water to prevent damage to annual or perennial crops.

Drier areas are often farmed by irrigation, and one would not consider drainage necessary. However, irrigation water always contains minerals and salts, which can be concentrated to toxic levels by evapotranspiration. Irrigated land may need periodic flushes with excessive irrigation water and drainage to control soil salinity.

Chapter 5

Arid-zone Hydrology and Water-Resources Development and Management

Concern over global climatic changes caused by growing atmospheric concentrations of carbon dioxide and so-called greenhouse gases has been a major issue of global environment and sustainable development. Recent symptomatic global climatic changes such as El Niño and the Sahel drought and/or desertification have made us more aware of atmospheric dynamics and global climate systems. Among the most important consequences of future changes in climate will be alterations in regional hydrological cycles and subsequent effects on the quality and quantity of regional water resources, yet these consequences are poorly understood.

Recent hydrological research strongly suggests that it is plausible that climatic changes caused by increases in greenhouse trace gas concentrations will alter the timing and magnitude of runoff and soil moisture, change lake levels and groundwater availability, and affect water quality. Such a scenario raises the possibility of dramatic environmental and socio-economic dislocations and has widespread implications for future water-resources planning and management.

Bv looking at the variations of climate and hydrological conditions in geological time, we can simulate the analogies of future greenhouse conditions which may cause significant changes in water availability. Nicholson and Flohn (1980), using a variety of palaeoclimatic records, suggested that there have been significant climatic changes in the Sahel region that can be related to possible driving forces in global atmospheric circulation, which may explain the huge amount of fossil groundwater resources stored in the extensive Nubian sandstones of the Sahara desert during the Late Pleistocene age.

Water differs markedly from most other natural resources by its remarkable property of continuous renewal in the water cycle, the main link in which is water exchange between oceans and the land. The world ocean is a gigantic evaporator, which, in this natural cycle, is the main source of fresh water. The fresh water falls as atmospheric precipitation and is the source of all water flows and water accumulation on land. The greater part of the water on the earth, approximately 97.5%, is salt water and water that is mineralized.

The volume of fresh water amounts to 35,029,000 km^3, or 2.52% of the total uncombined water on the earth. Rivers and streams account for 0.006%, fresh-water lakes for 0.26%, and water contained in the atmosphere for 0.001 % of the total quantity of fresh water. The rest of the fresh-water component occurs as soil moisture, permanent snow cover, marshes, and active groundwater.

Since the end of the nineteenth century the attention of many scientists has been directed to the problem of the connection between atmospheric precipitation, run-off, and evaporation in river basins. Water-resources development in the twentieth century has been directed to exploiting rivers and streams, lake water, and groundwater, which have had to be exclusively fresh. Saline waters such as seawater and brackish water, of which the storage potential is estimated to be as much as 97.5% of all the water on the earth, may suggest their increasingly important role in the water-resources planning of arid to semi-arid countries in the twenty-first century.

CAUSATION OF THE ARID ZONE

On a world scale there is a sharp fall in rainfall near 30° latitude in both hemispheres, and in places, notably on the west coast of Chile and Peru, rainfall is almost or entirely unknown. Since this aridity affects oceans as well as continents, it is apparent that there must be some process whereby rainfall is suppressed, for there is no shortage of water for evaporation. We shall look for this process among the dynamic factors associated with the general circulation of the atmosphere. The immediate causes of this aridity can be identified as follows: The first is the tendency towards divergent wind-flow at low levels, especially in the pole-ward parts of the trade winds.

This divergence causes a general subsidence of the air column throughout the low troposphere where the water vapour is concentrated. Dynamic warming associated with this subsidence tends to lower the relative humidity and disperse clouds; it also creates a significant degree of hydrostatic stability, so that convection currents and shower-type precipitation are inhibited.

At sea the trade winds have a shallow, moist layer of moving air capped by a stable or inversion layer above which the air is very dry. Over land, as in the Sahara, northern Mexico, Australia, and much of the Middle East (in summer), the dryness may extend to ground level.

Hence surface divergence, general subsidence, low humidities, and an absence of deep convection form a causally connected sequence over much (but not all) of the subtropical world.

The second cause of aridity, not truly independent of the first, is the existence of high-pressure zones near the latitudes of 30° north and south. Over the ocean, this high pressure is recorded at sea level, but inland it may be necessary to ascend to levels of 2-3 km to encounter the propagation of disturbances and the subtropical high-pressure belt, which is continuous in both hemispheres though considerably modified in form over southern Asia in summer.

These subtropical high-pressure belts also separate the circumpolar westerlies from the tropical easterlies. As is well known, both westerlies and easterlies are affected by travelling wave perturbations, which account for a good part of the precipitation in both regimes. The amplitude of both sets of waves is at a minimum near the axis of the subtropical highs; hence the subtropics are least likely to be affected by rain-bearing disturbances. This constitutes the second cause of widespread aridity.

In general it should be stressed that low humidities throughout a deep layer in the lower troposphere invariably lead to aridity, while very dry climates may also occur in areas of high atmospheric humidity. Thus parts of the arid south-western United States of America are dry not because of low humidities but because of the ineffectiveness of rain-making disturbances. A further point is that remoteness from the sea is not a guarantee of drought; high moisture contents occur deep in the interior of Amazonia, more than 2,000 km from the ocean along any possible direct streamline, and rainfall is heavy. Yet extremely low rainfalls occur in many areas along oceanic coasts, as in Chile, Peru, Morocco, and southwest Africa.

The point is, of course, that local sources of evaporation play a very small role in precipitation, which tends in most instances to fall from moist air streams whose humidity has been derived from very remote surfaces. The aridity of the subtropics thus emerges as an aspect of world climate dependent on deep-seated features of the earth's general atmospheric circulation.

It does not arise from local or man-made circumstances but from natural causes involving exceedingly large energy transformations and momentum transports. It has not been conceived that the regime can be significantly altered by human intervention. Discussion today in the early 1990s, however, is based on the worldwide concern with global climatic changes, which may be due to the substantial increases of carbon dioxide caused by industrialization since the Second World War.

It is equally unlikely that any past climatic epoch can have experienced a complete absence of subtropical aridity. As we have seen, the maintenance of the mid-latitude westerlies absolutely requires the existence of compensating easterlies in the tropics. Similarly the transfer of momentum and heat northward in the tropics requires the existence of a Hadley cell, with subsidence (and hence low humidity and drought) at some subtropical latitude. Hence it seems likely that the arid zone can have been no more than constricted in extent and driven a few degrees towards the equator at the height of recent glacial development; it can hardly have been eliminated altogether.

DEFINITION AND CLASSIFICATION OF THE ARID ZONE

Deserts have been mapped in regions as diverse as low and high latitude zones, inland continents, and coastal zones. Cold deserts, which are situated in high mountain ranges or high-latitude zones, are not included in this study, which deals with warm deserts situated between the tropics and the temperate zones. One of the simplest classifications of dry climates takes 10 inches (250 mm) as the dividing line between arid and semi-arid and 20 inches (500 mm) between semi-arid and humid. Although these criteria are scorned many scientific geographers and climatologists, they are actually not bad for the standard climatic classifications.

Most classifications today use combinations of temperature and precipitation, in order to make some allowance for the increasing evaporation with higher temperatures. De Martonne (1926, 1942) and Koppen (1923) both used figures of mean annual precipitation (P) and temperature (T). The basic de Martonne formula gives an index of aridity (I) which is a true sliding scale without artificial break points: $I = P/(T + 10)$. Koppen's formula has rigid break points, but these are varied according to the season of maximum rainfall. The margin between arid and semi-arid would be R (rainfall) = T + 11. Koppen's formulas were an attempt to assign climatic values to the limits of the main vegetation types of the world.

Another approach analyses the temperature and precipitation ratios of individual months. The monthly indices, including mean monthly precipitation (P_m) and temperature (T_m), are then summarized into an annual figure. Thus, de Martonne's index becomes the total of twelve monthly indices, each of which is calculated as $12P_m/(T_m + 10)$. Thornthwaite's basic system is more complex, but uses only mean monthly temperature and precipitation to arrive at a figure for estimated potential evapotranspiration (PEt) (Thornthwaite 1948). The arid regions are classified on the basis of an aridity index (AI) which assumes (1) AI = 0 when P = PEt, (2) AI = - 100 when P = 0, (3) AI = + 100 when P is much greater than PEt, with the following zone classifications corresponding to the AI ranges indicated:

- Semi-temperate, 0 to -10,
- Semi-arid, - 20 to -40,
- Arid, less than -40.

This method has been widely used by climatologists owing to its simple methodology. For the irrigation engineer, however, the method of estimating the parameters for potential evapotranspiration was found to be not accurate enough to be used for design purposes. Thornthwaite has introduced soil factors to the point that his latest indices can no longer be considered purely climatic. An early version of the concept of evaporation balance was devised by Albrecht Penck (1910), who used the geomorphic factor of hydrographic balance for a broad climatic classification. De Martonne (1942) drew a world map upon such a basis.

Emberger (1955) uses mean annual precipitation (P), mean daily maximum temperature of the warmest month (M), and mean daily minimum temperature of the coldest month (m), all combined into a single moisture quotient (Q) using the following formula:

$$Q = \frac{2P}{(M+m)(M-m)} \times 1{,}000$$

Unlike the formulae of Koppen, Thornthwaite, or de Martonne, Emberger's moisture quotient cannot be used by itself to make a valid climatic map. His maps appear to be based on his profound knowledge of vegetation, not on the mapping of climatic data. After mapping the vegetation zones, he determines the associated moisture quotients and other climatic values within the zones. Thus, the northern limit of his arid zone in north-west Africa varies from a moisture quotient of 16 to one of 40. In characterizing the climate of his stations, he uses the actual mean daily minimum temperature of the coldest month, as well as the moisture quotient.

His maps for north-west Africa, which he calls bioclimatic zone maps, are of fundamental value for all geographers and climatologists. Because of the accuracy and detail of his mapping, the maps form a valuable test of the vegetation validity of any climatic system. A simple check shows, for example, that de Martonne's index of 10 would serve fairly well as the northern limit of Emberger's Mediterranean arid bioclimatic zone.

Meigs, a former chairman of the Arid Zone Commission of the International Geographical Union, made a global arid zone map in 1951, whose zonal classifications have been widely used in arid-zone research since World War II. Meigs's classification is based on regional temperatures (mean monthly maxima and minima) and the duration of dry periods.

His arid index map distinguishes three subregions: extremely arid, arid, and semi-arid. The extremely arid region, which receives zero annual rainfall for twelve months of continuous observation, occupies about 4% of the land area of the earth. The arid region, which has a minimum of one month of rainy season, covers about 15% of the land area of the earth. The semi-arid region has a rainy season with 100-200 mm of rainfall and occupies about 14.6% of the land area of the earth.

GLOBAL ATMOSPHERIC CIRCULATION AND CLIMATIC CHANGES IN THE ARID ZONE

In the last twenty years low rainfall in the Sahel, possibly related to climatic changes, has reduced water availability in North Africa. Folland and Palmer (1986) noted that there is a strong inverse relationship between worldwide sea surface temperature (SST) and Sahelian rainfall. The SST has been rising since the late 1960s, which period corresponds with decreasing rainfall in the Sahelian region. The present increase in SST is considered likely to be linked to global climatic change, which suggests the warming of the climate by the greenhouse effect. However, the rising SST may also be linked to El Nino southern oscillations.

El Niño events occurred in 1972-73, 1976-77, and 1982-83, resulting in drought in the Sahel. Failure or weakening of the Guinea monsoon seriously reduced the rainfall in the Ethiopian highlands. Most of the major rivers of North Africa originate in the Central African uplands, the Atlas Ranges, and the Ethiopian highlands. More than 80% of the Nile's water originates in the Ethiopian highlands. The reservoir level in Lake Nasser has fallen by more than 18 meters in the last seven years owing to the African drought. As a result, usable effective storages are only onefifth (20×10^9 m^3) of the 1979 level.

The increasing albedo is another climatic factor that affects water resources. Desertification has increased the albedo level in the catchment area of northern African rivers. The net effect of a high albedo level is that the sun's energy is reflected to the atmosphere to heat the air, causing increased evaporation and transpiration, resulting in a decrease in the potential water resources.

PALAEOCLIMATOLOGY AND WATER-RESOURCES PLANNING

About 10,000 years B.P. the global climate was moist and cooler, indicating atmospheric carbon dioxide concentrations between 260 and 290 mg/l The earth was covered with 6.2 billion hectares of forest as compared to 4.1 billion hectares today at the end of the last (fourth) glacial age. Nicholson and Flohn (1980) explored water availability in Africa using a variety of palaeoclimatic records.

They suggest that parts of the Sahel region were drier 18,000 years before the present (B.P.), became moister and cooler with frequent rains in the period 10,000-4,500 years B.P., and then became drier again up to the present. Similarly, Goodfriend et al. (1986) explored the palaeoclimatic evidence for climatic changes in the area of the Jordan River basin and the Dead Sea. They identified large fluctuations in the level of the Dead Sea, its terminal lake, in the Late Pleistocene period up to 4,300 years B.P.

Reconstructions of the record back several centuries can provide valuable information on both past climatic conditions and the vulnerability of our water resources system to future changes. In a striking example, Stockton and Jacoby (1976) used tree rings to extend the run-off record in the Colorado River basin back more than 400 years, using carbon-13 (^{13}C) isotope in tree rings as the environment tracer. This kind of study has direct water management and policy implications. For example, the original 1922 Colorado River water allocation was based on the hydrological record available at that time: namely about 30 years from the late 1890s to the early 1920s.

In 1976, when the historical record was reconstructed back to the middle 1500s, the period from 1890 to the 1920s stood out as a time of abnormally high run-off. The 400-year record now shows that more water was allocated to users than is likely to be available on a long-term average basis. If the long-term record had been available in 1922, such over allocation might not have occurred. These changes can also be related to possible driving forces in the global atmospheric circulation.

PAEAEO-CARBON DIOXIDE AND THE GREENHOUSE EFFECT

The atmospheric carbon dioxide concentrations in air particles trapped in the 2,200 metre Vostock ice cores have been analysed to examine the historical changes of the climate since 160,000 years B.P. These Antarctic ice cores also provide temperature information for the same period based on the oxygen isotope (^{18}O) ratio. The derived temperature changes closely match changes in carbon dioxide concentrations. Gammon et al. (1985) reviewed the history of atmospheric carbon dioxide from 100 million years ago to the present. During the Cretaceous age (100 million years B.P) the carbon dioxide level was perhaps as high as several thousand parts per million; then it dropped to 200300 mg/l during the glacialinterglacial cycles of the past few million years through 10,000 years B.P.

Since the nineteenth century, the carbon dioxide level has increased to about 350 mg/l, which could be owing to the accelerated burning of fossil fuels as an energy source. North America, western Europe, and the former USSR and Soviet-bloc countries produce 67.4% of the world's carbon dioxide emissions to sustain their industrial activities. COHMAP AND GCMs.

The scientists of the Cooperative Holocene Mapping Project (COHMAP) have assembled a global array of welldocumented palaeoclimatic data and general circulation models (GCMs). The GCM, which was developed to look at soil moisture in the mid-continental region of the United States of America, has shown that significant drying may occur if the concentration of carbon dioxide in the earth's atmosphere is doubled.

GCM-generated hydrological data suffer from two major limitations: (1) the spatial resolution of GCMs is too coarse to provide hydrological information on a scale typically of interest to a hydrologist, and (2) hydrological parameterizations in GCMs are very simple and often do not provide the detailed information necessary for water-resources planning.

The hybrid COHMAP-GCM model simulates historical worldwide climatic changes in the atmosphere, geosphere, and biosphere that accompanied the transition from glacial to interglacial conditions during the past 18,000 years with geological and palaeo-ecological evidence, the results of which are to be used not for predicting the changes in water resources systems in future but for understanding changes in global climate on a geological time scale.

To assess the implications of the greenhouse effect for water resources in the arid to semi-arid regions, regional-scale details of future changes are needed for temperature, precipitation, evaporation, soil moisture, and other hydro-climatological variables. It is simply assumed that increasing concentrations of greenhouse gases will have an adverse effect on water resources in tropical and temperate arid zones.

Nature of the Hydrological Cycle and Global Water Balance

Hydrology is concerned with the occurrence and movement of water on the earth. Water is one of the commonest substances in nature. It occurs in chemically combined forms, in free states, and in the biosphere. The free states include groundwater and soil moisture in the upper layer of the lithosphere and in the soil cover, and water in oceans, seas, lakes, rivers, glaciers, and permanent snow cover.

A small amount of water occurs in the atmosphere as water vapour, water drops, and ice crystals. Unlike other natural resources, water is continually moving and changing from one form to another. The movement is a characteristic of all forms of water. The movements of ocean currents, rivers, groundwater runoff, humid air over oceans and continents, and transpiration are links in the interconnected water cycle of nature.

HYDROLOGICAL CYCLE

The rectangles of the figure denote various forms of water storage: in the atmosphere, on the surface of the ground, in the unsaturated soil moisture zone, in the groundwater reservoir below the water table, in the channel drainage network, and in the oceans. The arrows in the diagram denote the various hydrological processes responsible for the transfer of water from one form of storage to another. Thus the precipitable water (W) in the atmosphere may be transformed by precipitation (P) to water stored on the surface of the ground.

In the reverse direction, water may be transferred from the surface of the ground by evaporation (E) or from the unsaturated soil by transpiration through vegetation and subsequent evaporation from the leaf surface (Et). Some of the water on the surface of the ground will infiltrate through the surface into the unsaturated soil (F), but some of it may find its way as overland flow (Qo) into the channel network.

During precipitation, if the field moisture deficit of the soil which has arisen since the previous precipitation is substantially satisfied, then there will be either recharge (R) to the groundwater or else lateral interflow (Qi) through the saturated soil into the channel network. Groundwater storage is depleted by groundwater outflow (Qg), which enters the channel network and supplies streamflow during dry periods. During prolonged droughts, soil moisture may be replenished by capillary rise (C) from groundwater to the unsaturated zone and subsequent loss to the atmosphere by evapotranspiration.

Overland flow (Qo), lateral interflow (Qi), and groundwater outflow (Qg) are all combined and modified in the channel network to form the run-off (Ro) from the area for which the balance is being calculated. These various hydrological processes form the subject matter of physical hydrology.

GLOBAL WATER BALANCE

The quantities of water in the ocean, atmosphere, ice masses, lakes, and rivers may be evaluated without too much difficulty. It is more difficult to determine the amounts contained in living organisms and in the lithosphere. The figure of greatest uncertainty is that for inactive groundwater. The total reserves are preliminarily estimated to be about 1,386 million km. The breakdown of the volume of each form of water is shown below:

The volume of water in the ocean is estimated to be 1,338 million km^3 by assuming that the ocean covers an area of 361.3 million km^2 with an average depth of about 3.7 km. This represents about 96.5% of the total water reserves of the earth.

Table. World Water Reserves

	Volume (km^3)	Share of world reserves	
		Share of total water(%)	Share of fresh water(%)
Atmospheric water	12,900	0.001	0.01
Glaciers end permanent snow cover	24,064,000	1.74	68.7
Ground ice in zones of permafrost strata	300,000	0.022	0.86
Water in rivers	2,120	0.0002	0.006
Water in lakes (fresh, 91,000 km^3)	176,400	0.013	0.26
Water in marshes	11,470	0.0008	0.03
Soil moisture	16,500	0.001	0.05
Active groundwater (in aquifers), including brackish and fossil	10,530,000	0.76	30.1
Inactive groundwater (in lithosphere)	23,400,000	1.7	
World ocean	1,338,000,000	96.5	

Glaciers and Permanent Snow Cover

The amounts of water in the ice of the polar regions and in the glaciers of mountainous regions are estimated to be about 24 million km, which accounts for 68.7% of the earth's total resources of fresh water.

Inactive Groundwater

Substantial amounts of water are stored in the lithosphere. The inactive groundwater in the earth's crust-that is, gravity water contained in the pores and cracks of saturated strata-is estimated to be 23.4 million km³ by assuming an effective porosity of 5%-15% and a maximum depth of 2,000 m.

Active Groundwater

The depths of fresh-water accumulations vary, depending on local geological characteristics. By assuming an effective depth of aquifers between 200 and 600 m, the volume of active groundwater is preliminarily estimated to be 10.53 million km³, which is 30% of the total volume of fresh water.

Soil Moisture

Soil moisture is more closely related to weather conditions than is groundwater. During the wet seasons, moisture is stored in the soil, while it is removed by evaporation and tran spiration in the dry seasons. The storage of moisture in the soil is estimated to be 16,000 km³, by assuming a soil layer of 2 m thick with 10% of moisture on average.

Lakes and Reservoirs

There are numerous lakes in the world. Large lakes with an area of more than 100 km² may store 95% of the total reserves in all lakes in the world. The total water volume in the world's 145 large lakes amounts to 168,000 km³, which is 95% of the world's total of 176,000 km³. Of this, 91,000 km³ is in freshwater lakes, while approximately half the water (85,000 km³) is salty. Most of the salty lake water is concentrated in large lakes without outlet, such as the Dead Sea, the Caspian Sea, and the Aral Sea.

Intensive construction of large dams, especially since the Second World War, has created large reservoirs on major rivers. The total capacity of the 10,000 reservoirs of the world amounts to about 5,000 km³, with a net capacity of about 2,000 km³, which now controls approximately 14% of the total annual river run-off of 445,000 km³.

Marshes

Marshes occur in many areas of the earth. These are mostly peat marshes in countries with temperate climates and their equivalents in tropical and equatorial areas. The total amount of marsh water in the world is preliminarily estimated to be 11,000 km³.

River Channels

The total water storage in river channels of the world at any given moment is estimated at 2,000 km^3, which accounts for only 0.006% of all fresh water. However, it is of great importance to development as a renewable source for water supply.

Atmosphere

Water is contained in the atmosphere in the form of water vapour, water drops, and ice crystals. The total moisture in the atmosphere amounts to about 12,000 km^3, which is equivalent to 25 mm of water if spread over the whole surface of the globe.

Remarks on Global Water Reserves and Water Resources

The total amount of water in the hydrological cycle is constant and can neither be increased nor diminished. From the global-scale water budget study outlined above, it looks as though there is more than enough fresh water to meet the demands of human survival, both now and in the foreseeable future. However, water is often available in the wrong place, at the wrong time, or in the wrong quality. This uneven distribution is highlighted in the arid regions. In many of the countries in the Middle East, the hydrological cycle is being disturbed by overexploitation, depletion, or deterioration of the fresh waters in rivers and groundwater aquifers.

Salt waters, including seawater and brackish waters in rivers and aquifers, have been conceived as either useless or harmful and as being outside the scope of water-resources planning except where neither conventional river water nor groundwater of good quality exists.

The volume of conventional fresh-water reserves on the earth, however, is minimal compared with other forms of water such as seawater, which accounts for 96.5% of the total water on the earth. Potential groundwater reserves are estimated to be as high as 30% of the total freshwater reserves. However, brackish water is predominant in the major aquifers of the arid region owing to the minimal rainfall and hence minimal groundwater recharge.

Most of the known conventional water resources such as river water and groundwater of good quality or low salinity have already been developed or will soon be fully exploited in most countries of the Middle East. Non-conventional waters such as seawater and brackish waters, including both surface and groundwater, therefore, seem likely to play an increasingly important role in water-resources planning of the arid region for the twenty-first century, when the advances and innovations of desalting technologies, including reverse osmosis applications, are taken into account to save energy and cost.

THE INDUS RIVER

The Indus, one of the mightiest rivers of the world and the second longest in western Asia, has a mean annual discharge of 207.5×10^9 m^3. The Indus River system has ten times the volume of flow of the Colorado River in the United States and Mexico, and more than three times that of the Nile.

Alluvial plains in the middle to lower reaches of the Indus system have been developed to form the largest irrigation scheme in the world. Approximately 16,000 km of canals have been constructed to irrigate over 9 million ha. This irrigation project covers the greater part of a vast plain covered with fine-textured alluvial soil overlying coarser sediments extending deep (ten to hundreds of metres) downward to the bedrock of an ancient valley.

The development of the irrigation project was mostly carried out after 1850, but elements of an ancient flood irrigation channel can still be found. Most of the canals were excavated through the surface soil to the more pervious underlying fine sand, so that a large proportion of the surface water diverted through these canals seeped underground. The sediment, which forms an extensive aquifer, was continuously recharged by the leaking irrigation canals, so that the groundwater table rose continuously. In many areas, this rise was estimated to be approximately 30 cm per year. As a result, in much of the Indus plain the groundwater level was near the soil surface.

Proper aeration of the soil could not take place, and, with the prevailing arid climatic conditions, capillary action and evapotranspiration moved salts from the subsurface up to the root zone of crops and to the land surface. Thus once-fertile lands have so deteriorated that crops can no longer be grown. In the 1950s and 1960s, the salinity and waterlogging problems became so serious that intensive research was carried out by international agencies, headed by the Harvard University Water Resources Group, focusing on aquifer utilization and management, including the mining of groundwater. During successive stages of development of the irrigation systems in the Indus valley, little attention was given to land drainage, and emphasis was put on maximizing the extent of irrigated land. Integrated management of the underlying aquifer system in line with the surfacesubsurface drainage is now being used successfully to control salinity and reclaim irrigated land.

The River Basin

The Indus rises in Tibet, in the snow-clad Kailas range of the Himalayas, about 5,500 m above mean sea level. The catchment area extends over four countries-China, India, Pakistan, and Afghanistanwith the portion in Pakistan accounting for more than 50% of the total. The Indus basin lies in the subtropical zone; the Tropic of Cancer passes through its southernmost part, while its northern edge reaches the latitude 37°N.

The Indus cuts through mountain ranges forming a narrow gorge and deep channel from the headwaters until an important tributary, the Kabul River, joins it from the west near Attock. A few kilometres above the town of Mithankot, the Indus is joined by its most important tributary, the Panjnad River, which carries the waters of five main tributaries-the Jhelum, the Chenab, the Ravi, the Beas, and the Sutlej. The river slope from the headwaters to Attock is approximately 1/ 300; from Attock to Mithankot it is 1/4,000; and from Mithankot to the sea it averages 1/7,000. The total length of the Indus is about 2,900 km. The drainage area of the whole system is approximately 970,000 km^2.

The northern region of the basin, where the Indus and its main tributaries originate, is fully covered with rugged sky-high mountains and glaciers, comprising an area of about 452,000 km^2. Because of the steep and barren slopes of these mountains, the erosion is very heavy, ranging according to locality from about 400 to over 4,000 tons per km^2 annually, with an average of about 1,500 tons.

Lower down, the basin comprises vast plains formed and separated from each other by the five main tributaries. Further down from Mithankot, the Indus valley is covered with alluvium built up by the deposition of silt carried down by the river. The delta begins from Kotri, about 185 km from the sea, where the land is almost level and the soil is generally infertile. Closer to the sea is marshy land and mangrove forests which are generally flooded during high tides.

Hydrology

As noted, the Indus basin lies in the subtropical zone. In the plains, the average temperature during winter is about 21°C in Karachi and about 15°C in Lahore. Summer, the hottest season, is from May to August, and average temperature ranges from about 29°C in May to about 34°C towards the end of June or early July, when the maximum temperature often rises above 27.7°C. Despite the mighty Indus, the Indus plain is semi-arid. There are significant extremes of rainfall in the basin. The area around Sukkur and Mithankot receives only about 100 mm of rain a year, while, at Murree, a hill station at 2,280 m elevation, the annual precipitation is about 1,270 mm.

The precipitation, inclusive of snow, is many times heavier in the hilly region than in the plains. The mid-hill area, with elevations from about 1,200 m to 2,500 m, where the southwest monsoon generally strikes the mountain mass, has the heaviest rainfall, averaging about 1,250-1,500 mm a year.

At higher and lower elevations, the rainfall decreases rapidly and the air is correspondingly drier and clearer. The rainfall further decreases rapidly from north to south in the plains, from about 550 mm of annual rainfall in the foothills to only about 100 mm at Mithankot and Sukkur. Below Sukkur, the rainfall increases a little owing to the maritime air, reaching about 200 mm along the sea coast. The mean annual rainfall over the Indus plains is less than 250 mm.

Due to the uneven distribution of precipitation over the basin, the Indus and its tributaries receive most of their flow from the mountains. The flows are subject to extreme variations; the maximum summer discharge is over 100 times the winter minimum. During July and August, all the rivers attain their peaks, discharging a considerable volume of water to the sea. The mean annual run-off of the Indus is 207×10^9 m^3, with a yearly maximum of 264×10^9 m^3 and minimum of 171×10^9 m^3.

The Indus main stream carries about 110.3×10^9 m^3 per year, while the major tributaries Jhelum, Chenab, Beas, Sutlej, Ravi carry 27.85×10^9 m^3, 29×10^9 m^3, 15.65×10^9 m^3, 16.8×10^9 m^3, and 7.9×10^9 m^3 respectively. Thus the main stream alone carries a little more than half of the total discharge of the system. When combined with the Jhelum and Chenab, it carries a little more than fourthfifths of the overall total, while the Ravi, Beas, and Sutlej together deliver a little less than one fifth. The run-off coefficients are as high as 58%-82%.

Water-Resources Development

As rainfall is scarce in the plains, where the cultivable areas lie, agriculture on the Indus plains has to depend almost exclusively on an irrigation system utilizing the river flows. Many weir and canal systems were built on the Indus and its tributaries from the middle of the nineteenth century onwards, the first of which was the Upper Bari Doab canal, built between 1850 and 1859 to bring water from the Ravi River at Madhopur to the upper half of the doub, or inter-river land, in the vicinity of Lahore. By 1947, when India and Pakistan achieved independence as separate countries, the Indus water system had already been developed to provide irrigation to about 10.9 million ha, but it is remarkable that no storage reservoirs had yet been built in a system serving the world's largest irrigation area.

Among the major hydraulic works built before 1947 were the following:

- *On the Indus:* Paharpur canal, Sukkur barrage (North-West, D. Unhar, Begari, Ghoti canals).
- *On the Jhelum:* Upper Jhelum canal (Rasul barrage).
- *On the Chenab:* Upper Chenab canal (Marala barrage), Lower Chenab canal (Khanki barrage):
- *On the Jhelum and Chenab:* Trimmu barrage, Rangpur canal, HaveliSidhnai canal, Panjnad barrage and canal, Abasia canal;
- *On the Ravi:* Madhopur barrage (Upper and Central Bari Doab canals), Balloki barrage (Lower Bari Doab canal).
- *On the Sutlej and Beas:* Ferozepore barrage (Eastern and Bikaner canals), Suleimanki barrage (Pakpattan, Fordwah, and E. Sadifia canals), Islam barrage (Mails), Qaimpur, and Bahawal canals).
- *On the Swat:* Upper and Lower Swat canals.
- *On the Kabul:* Kabul River canal.

Because the new international boundary cut across the common canal system of Punjab, leaving one part in India and the other in Pakistan, controversy on the use of the canal waters arose soon after the Partition.

It took twelve years of patient negotiation before the controversy was settled by the Indus Waters Treaty in 1960. Between 1947 and 1960 intensive river development was carried out, including the following:

- *On the Indus:* Thal canals, Taunsa barrage (Dera Ghazi Khan and Muzaffargarh canals), Gudu barrage (Desert, D. Unhar, Begari, and Ghoti canals), Kotri barrage (Fuleli and Pinyari canals).
- *On the Jhelum:* Lower Jhelum canal (Rasul barrage).
- *On the Ravi:* Upper Bari Doab canal extensions.
- *On the Beas:* Shah Nehar and Hoshiarpur canals.
- *On the Sutlej and Beas:* Rajasthan canal.
- *On the Kabul:* Warsak dam.
- *On the Kurram:* Kurram Garhi weir.

The following major works were undertaken through the Indus Basin Development Fund (IBDF) and its successor, the Tarbela Development Fund:

- *On the Jhelum:* Mangia dam.
- *On the Jhelum, Chenab, Ravi, and Sutlej:* Trimmu-Sidhnai-MailsiBahawal link canal system, Rasul-Qadirabad-Balloki-Suleimanki link canal system.

- *On the Indus and Jhelum:* Chashma barrage, Chashma-Jhelum link canal.
- *On the Indus and Chenab:* Taunsa-Panjoad link canal;
- *On the Indus:* Tarbela dam.

It was originally intended that the IBDF would finance tube-well drainage works to compensate for leakage from the link canal systems, but at the choice of the Pakistan government the tube-well programme was financed by other means and the savings to the IBDF were put towards construction of the Tarbela dam.

Salinity and Waterlogging Problems

During successive stages of development of the irrigation systems, emphasis was put on maximizing the extent of irrigated land. In the Indus valley, as in all other flat valleys in the world, the natural surface and subsurface drainage is poor.

Since there were not enough drainage channels, most of the rainwater and canal seepage percolated down to lower depths.

As time passed, the groundwater table got higher and higher by steps, and finally, in the 1950s and 1960s it came close to the ground surface and has thus caused waterlogging in many large areas.

Proper aeration of the soil could not take place and the capillary action and evapotranspiration moved salts from the subsurface up to the root zone of the crops and to the land surface.

Thus, once-fertile lands so deteriorated that crops could no longer be grown.

In the former Punjab area in Pakistan, 5 million ha have already gone out of cultivation due to salinity caused by waterlogging, 690,000 ha are in an advanced stage of deterioration, and 2 million ha are affected to a lesser degree.

Since 1954, extensive groundwater and salinity investigations have been undertaken.

As a result of these investigations, recommendations were made to install a great number of deep wells to lower the groundwater level, and to use the pumped water for flushing the salts from the ground surface down to the drains provided as well as to the lower layer of soil.

Moreover, the pumped water is used to supplement the existing canal supplies for irrigation.

In 1959 the Water and Power Development Authority, established in 1958 to take charge of all water and power development in Pakistan, launched its first reclamation project in Rechna Doab, in the districts of Gujarawala, Sheikhupura, and Lyallpur. The project comprised about 1,800 tube wells in a gross commanded area of 480,000 ha. The average volume of groundwater pumped was about 1.85×10^9 m^3 per year.

In 1961 a second project was launched in Chai Doab, including 3,300 tube wells in a gross commanded area of 920,000 ha, with an average pumping volume of about 2.5×10^9 m^3 per year. The Authority has continued this salinity-control and reclamation work to remove the catastrophic threat to the well-being of the people of Pakistan.

Chapter 6

The Impact of Irrigation

Irrigation is generally defined as the application of water to the land for the purpose of supplying moisture essential to plant growth. In the process, irrigation projects transform the land in two ways:

- By direct modifications of the land surface that occur when canal net- works are constructed and land is cleared, shaped and levelled for irrigation;
- By indirect in-depth transformations that take place when the water and salt balances in the region are changed following the import of additional quantities of water and salt into the area.

Most of these transformations are permanent and irreversible. This chapter describes the dynamics of these transformations, and by using specific case studies examines the concomitant hazards that develop which may even threaten the existence of the society that created the irrigation project. Seasonal rainfall fluctuations make rain-fed farming.a risky venture. Irrigation reduces some of the uncertainties and so promotes increased production. In desert and arid zones, irrigation enables the economic development of areas which otherwise would be unproductive and uninhabitable.

Irrigation development is as old as recorded history. In about 2000 BC, an inscription on the tomb of the Assyrian Queen Semiramis proclaimed that 'I constrained the mighty river to flow according to my will and led its waters to fertilize lands that before had been barren and without inhabitants'. The Bible reports that the prophet Elisha in about 865 BC stated 'thus sayeth the Lord, Make this valley full of ditches... Ye shall not see wind, neither shall ye seen rain; yet that valley shall be filled with water, that Ye may drink, both ye and your cattle and your beasts'.

There is evidence of continuous irrigation for thousands of years in the Nile Valley and for comparatively long periods in Syria, Persia, India, Java and Italy. Egypt claims to have the world's oldest dam built about 5000 years ago to supply drinking water and for irrigation, and basin irrigation introduced at that time still plays a significant role in Egyptian agriculture.

Records show that irrigation in China was begun about 4000 years ago, and there are reservoirs in Sri Lanka more than 2000 years old. It is also interesting to note that the potential damages from irrigation have been known since early times. As far back as 2300 BC, the Babylonian Code of Khammurabi provided that 'If anyone opens his irrigation canals to let in water, but is careless and the water floods the fields of his neighbour, he shall measure out grain to the latter in proportion to the yield of the neighbouring field.'

An excellent historic example of extensive land transformations resulting from irrigation development is found in the stony-gravel limestone desert of the Negev area in Israel. Remnants of these ancient irrigation systems date back from the Israelite period (about 1000 Be) and from the Nabattean-Roman-Byzantine era (300 BC to 600 AD). In the absence of permanent water sources, the ancient farmers developed 'runoff' farm systems that used sporadic flash floods for irrigating.

Each farm unit comprised a watershed to collect runoff from the hillsides and a group of terraced fields in the valley bottoms for cultivation. The hillsides were subdivided into small catchments by channels leading the hillside runoff water to the terraced fields.

The stone and gravel cover was cleared from the ground surface of the catchments to increase the rates of runoff and placed in geometric patterned heaps of mounds and strips. Gravel mounds and strips, hillside collecting channels and stone-walled terraced fields transformed about 300 000 ha of the barren desert into a productive water-collecting system, and enabled the ancient settlers to cultivate about 15000 ha and establish a permanent agricultural civilization in a 100 mm annual rainfall region.

THE PRESENT AND FUTURE EXTENT OF IRRIGATION

The area of land irrigated in the world today is close to 250 million hectares, of which about two-thirds is situated in five countries-China, India, Pakis- tan, the Soviet Union and the USA. With world populations growing concur- rently with expectations of higher standards of living, world food and fibre demands will increase.

Since arid lands become highly productive with the introduction of irrigation water, and the productivity of rain-fed areas increases considerably with supplementary irrigation, the improvement of existing irrigation projects and the opening up of new irrigated areas will continue to make significant contributions to solving the world's expanding food and fibre needs. In order to ensure the stability and permanency of these projects we cannot afford to overlook the potential hazards intrinsic to irrigation development, and we must be aware of the land transformations that may take place.

LAND TRANSFORMATIONS

Land transformations resulting from irrigation development are in all cases spectacular-especially when seen from the air or recorded in satellite imagery. Aerial photographs showing the boundaries between the desert and the sown in, for example, the Nile Valley in Egypt, the Central Valley in California, or the Rajasthan Canal in India, exhibit dramatic pictures of the contrasting land surfaces and highlight the transformations that have taken place as a result of irrigation making the deserts blossom.

These land transformations can be classified into two categories:

- Surface modifications resulting from implementing various irrigation methods.
- In-depth changes caused by disturbing the local or regional salt and water balance.

Most of these transformations are irreversible and permanent. All irrigation projects involve transferring water from one area to another-from a water- surplus region to a water-deficient one. Some major irrigation schemes call for inter-regional water transfers from one basin to another using major dams for inter-annual or inter-seasonal storage. They often require main canals bundreds of kilometres in length to transport the water.

Inherent in this irrigation development are the land transformations caused by the construc- tion of these large multi-purpose dams-such as those that have followed the building of the Aswan, Kariba, Tarbella and Boulder Dams. Reservoir sedimentation, steam bed aggradation, flooding of upstream forests and villages are only some of the many problems concomitant to major schemes. This chapter, however, is confined to land transformations that are directly related to the irrigation development.

SURFACE TRANSFORMATIONS

Irrigation water is applied to land using one of the four main methods:

- Surface flooding.
- Partial wetting of the surface in furrows.
- Surface sprinkling.
- By drip (or 'trickle').

All methods, particularly surface ones, involve considerable land preparation. The first land transformation follows clearing the land of trees, brush or other vegetation. In the semi-arid zones (for example in the western part of the United States) relatively dense cover of sagebrush, grease-wood, chapar- ral, scrub oak, pine or mesquite have to be eradicated from proposed irrigated lands, and in the monsoon areas such as Thailand, India, Indonesia etc. tropical and sub-tropical jungle forests must be cut down and eliminated before development can begin.

Small brush up to 0.5 m in height can be removed by ploughing, while vigorous vegetative growth usually requires the clearing operations to be carried out with large and powerful tractors. In some cases deep-rooted tree stumps must be blasted away-leaving the land with a bomb-pitted look of a war zone. In other cases stones, boulders and rock outcrops have to be hauled away from the project lands.

The land having been cleared, it next becomes necessary to smooth (or grade the surface) to facilitate uniform water distribution. These works bring about another land transformation. Surface smoothing is usually accomplished by the use of harrows, graders, scrapers, bulldozers and land planes- but in under-developed countries hand labour and bullock-drawn wooden planning boards are often used for this work.

On accurately leveled land, water will be distributed uniformly over the fields; but when the leveling is done poorly and the ground surface remains rough and irregular, some parts of the fields will receive more water than others. The final shape of the land surface will depend on the method of irrigation to be used. In many of the older irrigation systems, where the wild-flooding method is common, water flows from the supply ditches over the fields and is guided only by the slope of the land.

This method requires minimum amounts of land leveling (say, up to about 200 cubic meters of earth moving per hectare), but irrigation by this method is generally the least efficient of all methods. More efficient methods of surface flooding call for accurate land shaping and grading and 1500 cubic meters of earth may seen to be moved per hectare.

In the border method of irrigation the land is divided into long, narrow strips 5-15 m wide, extending for 100-400 m down the prevailing land slope, and the border strips are separated by levees, 0.1-0.3 m high. Border strips are leveled transversely, so that water will flow at an even depth per the whole width. The longitudinal slope is graded to a design gradient to enable an even percolation of the irrigation water over the entire border.

In the level basin method the land is shaped into accurately leveled plots ranging in size from 0.2 to 1.0 ha and surrounded by low bunds. The basin method is suited both to permeable soils which must be covered quickly with water in order to prevent excessive losses near the supply ditches, and to heavy soils with slow percolation rates. It is the method most commonly used for rice ('paddy') irrigation all over the world.

Small, level basins are also used for irrigating orchards. Leveling works for constructing either borders or basins transform the original natural land surface into clearly discernable rectangular or square patterned fields. In the irrigation methods described above, almost the entire land surface is wetted at the time of irrigation. When furrow irrigation is used, the fields are only partially wetted.

Furrows (20-30 cm in depth) - or corrugations (10-15 cm) are spaced according to the plant rows-one furrow or corrugation being provided for each row. For efficient furrow irrigation the land must be graded accurately and the flow of water into the furrows controlled. The pattern created by this method is neither square or rectangular, but the fields are formed into long, thin, parallel linear strips leading away from the delivery ditches.

In sprinkler irrigation projects, water is applied to the crop in the form of a spray. Unlike the surface methods, sprinkler irrigation requires less land leveling and can be used economically on undulating lands, on soils too shallow to prevent proper leveling, and on soils too porous for good water distribution by surface methods. The land transformation process is partly dependent on whether the sprinkler system is semi-permanent or portable.

A particularly dramatic land metamorphosis takes place when central pivot installations are introduced on a large scale. In these schemes, combinations of rotating and fixed sprinklers deliver water from central pivots through continually rotating booms to circular-shaped fields. The fields can have diameters of up to 1 km. These systems when seen from the air or on satellite photographs show up as green circles in a patterned design on a beige desert background. Drip irrigation systems consist of extensive networks of small-diameter plastic pipes delivering low flows of a few litres per hour to the plants through special emitters. From each emitter, water spreads laterally and vertically in the soil profile.

Drip systems require little land leveling. Dissolved salts are left in the soil near the edge of the wetted zone and these salts must be periodically leached out of the root zone by rainfall or by irrigation. The system is adaptable to a wide variety of shallow, sandy or stoney soils as well as undulating and steep topographic conditions. With the world's expanding population pressures, planners will be forced to bring the more difficult areas into production, and we can expect that extensive areas will come under drip irrigation in the future to enable otherwise unirrigable areas to be developed.

In-depth Transformations

Since man first began to question the nature of the four basic elements (water, fire, earth and air) he has pondered on the behaviour of water. The early philosophers were puzzled by the continual flow of rivers to the sea and wondered why the level of seas did not continually rise. The early Greek philosophers of the fourth and sixth centuries BC imagined immense under- ground reservoirs from which the rivers extruded, and it was not until the advent of the early Roman philosophers in the first century AD that the 'evaporation-condensation' phase of' the hydrological cycle came to be understood.

This lack of appreciation of the hydrological processes contri- buted to the disintegration of some of the ancient irrigation systems. It was not until the Middle Ages, however, that da Vinci gave the first realistic representation of the water balance equation, when he wrote that in the course of a year, the amount of water that rises from a particular region will be as great as the flows in the rivers and the rainfall that descends in that region.

In the nineteenth century the hydrological cycle was first analyzed as a system of storage with a given input producing a specific output. The quantity of water moving through the hydrological cycle in a unit of time could then be evaluated from the water balance equation:

$$J\text{-}O=S$$

where 'I' is the inflow during a given period (surface water, groundwater and precipitation); '0' is the outflow during the period (evaporation, transpiration, surface runoff and groundwater discharge from the area); and 'S' is the change in storage in retention, depression, interception, soil moisture and groundwater levels. Implicit in this approach is the understanding that feed- back mechanisms in the system constantly tend to bring about some state of equilibrium.

In a non-irrigated area a stable but changing water balance is developed between the annual rainfall input, ranging from 100 to 1000 mm, the vegetation and the groundwater. With the introduction of irrigation the balance is disturbed. Since irrigation may add the equivalent of, say, 500-1500 mm of water to the annual rainfall, this three- to ten-fold increase in water input will significantly upset the previous prevailing water balance.

From its original steady state of equilibrium with system variables oscillating around long-term averages, the additional input forces the entire system over some threshold value into new levels of equilibrium or into meta-stable equilibrium states. The rate of change to the new levels will depend on the pedology, geology, climate and crops grown in the region.

It is important to recognize all the changes that can result from human intervention into the input-output response-process and how the modified hydrological system transforms the land of a particular region. The transformations will depend on both the spatial and the time scales over which the intervention is applied. A few in-depth micro-scale transformations may be observed in relatively short time periods measured perhaps in years, but most macro-scale changes require decades or even centuries to become hazardous threats to the permanency of the system.

Some effects may be local and cover only a few hectares, while others may be regional and affect thousands of hectares. In projects where irrigation is supplemental rather than continuous, these effects are generally limited and easier to control. This chapter focuses on the large-scale transformations that develop as dynamic meta-stable processes passing through a number of thresholds before they become problematic and approach a state of irreversibility.

Four deleterious land transformations follow irrigation development:

- Waterlogging.
- Saline and alkaline soil development.
- Sea-water intrusion.
- Land subsidence.

These transformations are now described and illustrated with some typical case studies.

WATERLOGGING

Waterlogging is the result of a rise in the groundwater level. An area is generally classed as waterlogged if the water table lies less than 2 m from the ground surface. When waterlogging reaches the root zone of the plants, yields diminish significantly because the roots need a soil-air-water environment to grow and cannot survive in free water. There are four principal causes of waterlogging-leakage from canals, wastage from distribution networks, over-irrigation and lack of suitable drainage facilities.

Following the water- logging process, saline conditions generally develop on the surface. Many examples of waterlogged areas can be found in the irrigated areas of the world-in India, Pakistan, the USA, Australia, North Africa and China. The case history of Pakistan has been selected because it is a good example of an extensive occurrence of the phenomena and demonstrates the engineering and management measures governments must introduce in order to control regional waterlogging.

There is evidence that irrigation has been practised along the Indus River system in Pakistan from 3000 BC onwards. In the beginning, only narrow strips of land adjoining the river banks were irrigated. With time, irrigation has been extended. Pakistan contains probably the largest irrigation system in the world, and today about 15 million hectares are irrigated, of which more than 75% is served by government canals. The irrigation system comprises the Indus river and its major tributaries, 3 storage reservoirs, 19 barrage/ headworks, 43 canal commands and about 90 000 watercourse systems. The total length of the canals is about 56 000 km and there are more than 1.6 million kilometres of watercourses and farm channels.

The physical layout of the canal systems and the methods of delivering water were evolved to fit the pattern of available supplies in the unregulated rivers, and to conform to the legal conditions and water allocations of the Pakistan-India Indus Water Treaty of 1960. The system was planned to bring to maturity the largest possible cropped area with the minimum consumption of water. Consequently, irrigation intensities are generally low (25% in the kharif season and 50% in the rabi season) so that the water can be spread over as large an area as possible.

A basic canal design concept underlying the planning has been to achieve equitable water distribution with the minimum of human interference in the management of the system. Overall irrigation efficiencies, however, are relatively low-possibly about 30%-and studies show that 35-45% of the water delivered by the canals never reaches the farmers' fields. Since the natural drainage is inadequate and artificial drainage was not provided in the early project years, extensive over-irrigation combined with excessive leakage from the canals have led to a rise in the water table.

Rates of rise of the water table have been relatively rapid-about 30-40 cm a year. Depending on the initial depth of the water table, areas along the main canals were the first to become waterlogged, but subsequently waterlogging spread to the contiguous areas. The potential need for drainage to prevent this land transformation was realized when large-scale develop- ment was initiated, and a network of observation points was established to monitor the changing groundwater conditions. Waterlogging conditions measured in 1978 showed that in the 14.6 million hectares surveyed, the water table was less than 1.8 m below the surface on 13% of the area and between 1.8 m and 3.0 m on 41% of the area.

In some areas, the water table is still rising; but in those projects where the government has initiated drainage schemes, the water table is slowly being controlled and some cases even lowered. Unless these drainage measures continue to be implemented, 50% of the irrigated fields will have to be abandoned in the future and the land will be transformed into unproduc- tive saline waterlogged swamps, which will be useless to Pakistan.

SALINITY-THE SECOND OF THE 'TWIN MENACES'

All irrigation water contain salts and the salt concentrations in soils tend to increase as water evaporates from the surface or is transpired by plants. In order to maintain a favourable root-zone salt balance, more water (called the 'leaching requirement') must be applied to the soil tan is used by the evapo-transpiration process. The excess water drains to the groundwater table, and in the absence of drainage it contributes to the development of waterlogging conditions described in the foregoing.

Salts accumulate in the upper soil profile as evaporation and transpiration remove water, and higWy saline or alkaline conditions develop as more and more water moves upward from the water table by capillary action. This salinization of the upper soil profile increases continuously and when it reaches intolerable levels the land goes out of production.

The rate of salt accumulation depends on the soil surface temperatures, the chemical composition of the ground and surface waters and on the physical and chemical properties of the soil. As the balance between the sodium, magnesium and calcium ions changes in the soil, sodium becomes absorbed on the fine clay particles and the soil properties are modified. Soil swelling, colloidal dispersion and void clogging create impermeable soil profiles and unsatisfactory root aeration conditions, with subsequent significant reductions in production.

In order to prevent the development of these salinity conditions in irrigated areas, two courses of action are possible:

- Installing surface or subsoil drainage to remove water from the upper 2 or 3 metres.
- Developing vertical drains (wells) to lower the water table with the use of pumps.

The first method is feasible where suitable outlets for the drains are available. However, this solution may only transfer the problem downstream because when saline drainage water is discharged into local rivers, irrigation projects situated further downstream may be adversely affected. For example, the Colorado River's salinity has been rising during the last few decades as it flows downstream through the States of Nevada, Arizona and California.

The current level of salinity at the Imperial Dam is about 900 ppm (parts per million), and is predicted to continue to rise and to reach about 1200 ppm in the year 2000. The cost of the damages due to this high saline content is expected to amount to about $80 million a year and indicates the stake that the irrigators and urban water users in the Colorado Basin have in the measures needed to control the salinity.

The second method of controlling water levels in the ground, pumping from wells, requires a relatively costly energy input to operate, but it has the advantage that a portion of the groundwater pumped can be used for irrigation in conjunction with the surface water. The management of a conjunctive operation, however, can only be done on a large scale and often requires government control of the regional water balance.

Pakistan also provides an example of the salinity problem. As mentioned previously, large-scale irrigation development in Pakistan led to rapid rises in the groundwater table. Groundwater has been a traditional source of irrigation water in the region, and the predominant method of using groundwater was, until the 1970s, the 'Persian-wheel' driven by draft animals. Over 200 000 units are still reported to be in operation. During the last 20 years, modern diesel- and electric-powered tube wells have been introduced in both the public and private sectors.

The water-salt balance in the Indus Basin represents a complex and difficult problem because the average slope of the area is about 1:5000 and there is no natural drainage for disposing of saline effluents. The irrigated plain is underlain by an extensive groundwater aquifer covering about 16 million hectares. One-third of this area has 'fresh' groundwater (1000 ppm TDS-total dissolved salts), one-tenth a moderate salinity content (1000-3000 ppm TDS) and the remainder is highly saline (>3000 ppm TDS). Total recharge of the aquifer is about 50-75 x 109 cubic metres per year.

At present about 180000 private tube wells and 12500 public tube wells jointly pump about 45 × 109 cubic metres per year from the ground- water. Without this pumping, the Indus Basin would be a large lake. The Indus Basin has also been acting as a large salt sink. Average annual river water diverted for irrigation is about 100-120 × 109 cubic metres per year, with an initial average salinity of 130 ppm. This means an annual addition of 16 million tonnes of salt to the 16 million hectares of irrigated land, i.e. about 1 tonne of additional salt is added annually to every hectare. With the recognition of the expanding salinity hazard in the late 1950s, the first large-scale groundwater development and salinity control programme was initiated with the Government Salinity Control and Regulation Project (SCARP I).

The plan proposed to control salinity by pumping large quantities of water from the aquifer, spreading if over the land and so leaching down the salts. Parallel with the SCARP I programme, private tube well development took place at an accelerated rate and total private groundwater development today far exceeds the public tube well programme. Because of its success, SCARP I has been followed by a series of five-year SCARP plans and SCARP V is currently in operation.

The design of the SCARP projects is based on sophisticated criteria for handling the difficult groundwater and aquifer conditions, such as delineating mixing zones for conjunctive use of different levels of groundwater and surface water, and the programme includes mining certain groundwater areas, combining private and public tube well development, and remodelling canals to increase surface water supplies.

The SCARP programmes are at present under review and an overall strategy is being developed to create a permanent water-salt balance in the Indus Basin by adopting strategies for both prevention and cure. Plans are now being evaluated to dispose of the salts by:

- Constructing large outfall drains to discharge highly saline effluents to the lower reaches of the Indus River and to the sea during the flood season.
- Transporting some of the saline water to large evaporating areas in the desert.

During the past 20 years, 30-50% of the water sector's budgetary allocations have been spent on implementing salinity control measures. In the future similar large expenditures will have to continue to be invested in eliminating the long-term salinity hazard of the Indus Basin.

SEAWATER INTRUSION

Two types of seawater intrusion will be discussed: (a) subsurface intrusion into coastal aquifers; and (b) surface flow intrusion into rivers.

(a) Subsurface coastal sea intrusion

Seawater intrusion into coastal aquifers is a problem of worldwide significance. It results when excessive groundwater pumping causes lateral and upward movement of seawater into the coastal wells. Unless controlled, the phenomenon causes significant rises in the salinity levels of the coastal groundwaters and concomitant detrimental effects on the lands being irrigated by these sources.

Coastal aquifers discharge their unexploited groundwater to the sea. In many countries these aquifers serve as major water supply sources for the intensively developed agricultural areas and densely settled municipal areas in the coastal plains-such as in California, Florida and Israel.

Exploitation of a coastal aquifer can be maintained indefinitely without causing hazardous consequences provided the maximum 'safe yield' of the aquifer is not exceeded. Simulation models for the behaviour of coastal aquifers have been developed in many countries to evaluate potential yields and are useful tools in predicting the behaviour of the groundwater system under alternative abstraction options.

These models, although using idealized conditions and approximate assumptions, give results sufficiently accurate to enable the planner to establish the rate of groundwater development most suited to the local or regional conditions. Since lateral and vertical flow rates of groundwater are extremely slow, a critical lowering of the water table near the coast does not bring on immediate seawater intrusion because of the relatively long period required for the salt water to move inland. Thus in interpreting groundwater model results, it is extremely important to give full attention to the time scale which will generally be measured in decades rather than in years.

An excellent sample case history of successful management to date of a coastal aquifer and the prevention of the damaging effects of seawater intrusion on a regional scale is found in Israel. Israel's main water system is a nationwide grid of interconnected waterworks in which surface, ground and reclaimed water are integrated. Israel's water supply is based mainly on two elements:

- The waters of the Jordan river and of Lake Kinnereth in the North.
- The groundwaters of two principal aquifers-the coastal sandstones and the Yarkon-Turonian limestones.

Other than Lake Kinnereth, there are no favourable surface storage facilities in Israel and the need to provide regulation falls on the two main aquifers. These aquifers make up about 60% of all the nation's freshwater resources, and the coastal aquifer which supplies about 30% of the water consumption of Israel is used as a major long-term regulating reservoir.

This aquifer flows towards the Mediterranean, is phreatic with water level depths ranging from a few metres to about 90 m below the ground surface. The aquifer is replenished by percolating rainfall and return flows from irrigation and waste water. The coastal plain overlying the aquifer has a total area of about 2 million hectares and includes the major urban communities of Israel, account- ing for about 50% of Israel's population (in 1983 about 3.3 million). The projected population of the coastal plain is expected to double by the year 2010.

The coastal aquifer has an estimated yield of about 160 x 106cubic metres a year and the limestone aquifer about 200 x 106 cubic metres a year. In order to enable large-scale immigration into Israel and a rapid development of agriculture during the years 1955-64, the coastal aquifer was exploited at an average rate of 330 x 106cubic metres a year and the limestone aquifer at a rate of 260 x 106cubic metres a year, causing an annual overdraft amounting to about 230 × 106 cubic metres.

Over-pumping of the groundwater led to serious intrusion of seawater into the heavily over-exploited coastal aquifer. By the year 1964, however, the National Jordan conduit had been completed and water could be brought from Lake Kinnereth in the north, to supply the needs of the south. Some of this imported water was injected into wells in the coastal plain in order to repair the saline intrusion damage as rapidly as possible. Since then, using the results of simulation models, groundwater abstraction has been controlled and reduced to rates that ensure a sustained long-term programme of conjunctive use of Jordan water and groundwater.

The combination of a shallow phreatic water table, sandy soils, dense populations, intensive agriculture, and excessive pumping could have led to a potential disaster in Israel. Sound long-term planning using simulation and hybrid computer models, backed up by suitable water laws and competent government management, enabled the full use of the water resources to be achieved without creating critical irreversible and disastrous conditions.

Prob- lems still remain. Water quality records show that considerable parts of the aquifer are polluted by nitrates, and unless this is controlled the quality of a major portion of the urban water supply wells in the coastal aquifer will exceed the maximum allowable drinking water standards by the year 2000. On-going studies are evaluating the appropriate engineering techniques needed in the coming decades to ensure that the quality of the aquifer water conforms to these standards.

Seawater intrusion into rivers:

As more and more fresh water supplies are diverted or pumped from rivers, the flows reaching the sea decrease; eventually a point is reached when tidal flows begin to move upstream and cause deleterious increases in river salinity. This problem is worldwide, and many downstream riparian areas suffering from serious salt damage will go out of production unless a timely solution is found. The case history of the Chao Phya river in Thailand illustrates the problem.

The Chao Phya provides the major water source for the irrigation in the Central Plain and Delta of Thailand. The average annual supply is about 23 × 109 cubic metres a year, although variations of flow from year to year may be large. Flows reaching the Central Plain are being gradually reduced owing to intensification of rice production in the Northern Regions and expanding irrigation development along some of the upstream tributaries of the Chao Phya.

The total crop land in the Central Plain is about 1.4 million hectares in the wet season, and the Royal Irrigation Department (RID) plans to increase production by expanding dry season cultivation from about the present 0.2 million hectares to more than 0.6 million hectares. The additional water for this development must come from storage in upstream reservoirs on the Chao Phya and its tributaries. This will aggravate the present hazardous condition when, during periods of low flow in the dry season (January to May), seawater flows upstream for about 30-40 km from the Chao Phya estuary.

The irrigated areas affected today by the saline water intrusion cover about 1500 km^2, and the seawater intrusions also have deleterious effects on the domestic water supplies. The problem is a difficult one to solve because of the conflicting interests between the developments targets of the Northern Regions and of the Central Plain. It has been estimated that a minimum flow of about 150 cubic metres a second must be maintained in the Chao P.hya river at Bankok to ensure a salinity level of about 1000 ppm (TDS). However, 30 m3/s of this flow will eventually be required to supply water to the Bankok area, and only 120 m3/swill remain to flow south of Bankok. Under these conditions the salinity level will rise to 2000 ppm (TDS).

A study made by the RID in 1968 concluded that the value of water released from the upstream dams was greater for salinity control than for power. It is essential to assess accurately the amount of water to be allocated to control seawater in the Chao Phya river and to supply water to all the available irrigable land. Hence priorities for development must be established on a basin-wide plan to take into account irrigation, power, municipal supplies and salinity control requirements.

LAND SUBSIDENCE

In recent years it has become apparent that extensive exploitation of groundwater has brought about land subsidence in many localities of the world. A casebook being prepared by UNESCO reports on the occurrence of more than 40 major land subsidence areas in the world and also describes methods for measuring, predicting and controlling the phenomenon. Subsidence due to excessive groundwater pumping develops under two geological conditions.

First, in carbonate rocks overlain by unconsolidated deposits that receive buoyant support from the groundwater, when the water table is lowered that support is lost; unconsolidated material may move downwards into openings in the carbonate rocks, often causing 'catastrophic collapses.

Second, in semi-consolidated sediments of high porosity underlain by sand-gravel aquifers of low compressibility and inter-bedded with clayey aquitards of low permeability and low compressibility, pumpage from the aquifers is accompanied by vertical leakage from the aquifards.

Since the compressibility of the clay is one or two orders of magnitude greater than the compressibility of the sand, the total compaction of the clay aquitard is much greater than that of the sand-gravel aquifer; the sediments undergo irreversible compaction as water is squeezed out of the interstices and result in land subsidences. Since the hydraulic conductivity of the clay is several orders of magnitude less than that of sand, the subsidence process is a slow one.

Land subsidence ranges from as little as 20-30 cm to up to 900 cm (in some parts of California), and the areal extent from small areas 10 km^2 in size to extensive zones covering up to 13500 km^2 (also in California). The principal problems arising from land subsidence are:

- Differential changes in the gradients of streams, drains and canals.
- Failure of wells due to rupturing of the well casing.
- Tidal encroachment in lowland coastal areas.
- Damage to buildings, roads, railways, power lines and water supply systems.

Land subsidence in Arizona is used here as a typical case history. Arizona's current consumption of water from all sources is about 5.9 × 109cubic metres a year, of which only about 3.4 × 109 cubic metres are renewable-hence Arizona is mining its groundwater at almost twice the natural replenishment rate, and during the last 30 years Arizona has already overdrawn about 75 × 109 cubic metres.

The main culprit for this rate of depletion is agriculture, which uses about 90% of all water consumed in the state. The only way to solve this problem without radically reducing the irrigated areas is to bring in additional water supplies. This is the purpose of the Central Valley Project which, when completed, will include a 480 km canal transporting annually about 1.5 × 109 cubic metres of water from the Colorado River into Arizona.

This, however, will eliminate only part of the overdraft. Consequently, the State Legislature has passed the Groundwater Act of 1980 to maintain a long-term balance between groundwater withdrawals and recharge (by natural or artificial means). Arizona has become the first state in the USA to limit the pumping of groundwater. In order to achieve an eventual balance, it is already recognized that agriculture's share of the state's water resources will have to be reduced from 90% to about 70%. This will be accomplished by changing present cropping patterns, improving irrigation efficiencies, reduc- ing waste, introducing more efficient water consumption crops, and initiating a recycling programme for use of sewage effluents.

MODERN PLANNING METHODOLOGIES

Concern with the relationship between land transformation and irrigation is not new. In Mesopotamia, for example, where irrigation was developed on a regional scale in about 2300 BC, the country faced serious salinization difficulties-the earliest and probably most serious land transformation occurring at about 2000 BC when many settlements had to be abandoned. Since that time, irrigation planning has made tremendous strides.

Engineers have learned the lessons of earlier civilizations, and today modern technology gives the engineer techniques to analyse the problems and predict the behaviour of the water system on both macro and micro scales. The tools are available to avoid a number of mistakes of the past.

Man's use of the soil has advanced through three stages. In the first stage, an ecological balance was maintained because shifting cultivation, or low irrigation intensities, did not deplete the soil fertility permanently. In the second stage, soil exhaustion occurred because the fertility of the soil was mined without replacing nutrients removed by cultivation, leaching or erosion. In the third stage, man has learned to conserve the productivity of the soil by returning as much or more than he removed, so that permanent cultivation or irrigation has become possible.

The objective in irrigation management must be to maintain a dynamic balance in the salt-water input-output relationship and to replace fertility losses with fertilizers and with good farm management. There is no need to assume that technology will remain constant, nor need we prophesy doom for modern irrigation projects.

Historical evidence favours the view that 'necessity is the mother of invention', and new technologies will be developed to meet any potential hazard that may arise-particularly since these problems do not appear suddenly but develop over a decade or two. A good example of present-day regional irrigation planning can be found in the Rio Colorado Basin Project, the plan for which was prepared in the 1970s and included more than 20 irrigation sub-projects covering a total of about 0.8 million hectares.

This basin-wide study used multiple social, economic and regional objectives to evaluate various development options in an overall water resource plan, and developed mathematical simulation and optimization models to predict potential problems, future benefits and investment costs. A more recent major study is being carried out in India on a scheme funded by the World Bank-the Sadar Sarovar Project in the State of Gujarat.

This multiple- purpose hydro-power and irrigation project will use the waters of the Narmada river to irrigate 2.2 million hectares of land from the year 1990. The Narmada Planning Board established by the State of Gujarat has developed optimization models to examine the benefits from alternative cropping pro- grammes and, because groundwater and salinity problems are expected to develop, surface and groundwater simulation models have been set up to predict the rates of groundwater rise in various zones of the project.

The results of these simulation models are already being incorporated into the design, and extensive public and private tube well schemes for conjunctive use of ground and surface waters are being anticipated so as to esure the permanency of the project by maintaining a satisfactory salt-water balance in the region.

STABILITY AND PERMANENCE

Surface and in-depth land transformation always follow in the wake of irrigation development. The surface modifications take place during the construction period, while the in-depth changes appear only after the intro- duction of water into an area. While surface modifications can be regarded as beneficial to man, the in-depth changes carry inherent potentially harmful features-waterlogging, salinity, seawater intrusion and land subsidence. These hazards must be controlled in time lest their deleterious effects become malignant and reach dangerous and irreversible levels. Good management is therefore a key ingredient of irrigation development. Management must continually monitor the water and salt balances so as to predict the rates of build-up of the hazards. Finances must be made available in time to introduce the measures needed to control the changes so that they do not reach injurious states, placing the project in jeopardy.

The main lessons to be learned from the history of irrigation are that governments must recognize these potential hazards, in the early project planning stages, and establish competent operational management structures to detect and heed early-warning signals that precede problematic situations. There are no perfect solutions, and choices will always have to be made between various policy options, trading off losses in production and farmer's income in order to achieve long-term stability and permanence.

Good planning, the quintessence of engineering, is essential to ensure lasting survival of an irrigation project and the society dependent on it. In the past, emphasis in planning has been placed on the need to meet plant water demands to attain maximum yields (per unit of land). Greater weight should now be given to the management of supplies, with the aim of maximizing economic returns on water (rather than land) and minimizing long-term hazards.

In carrying out the economic analysis of a project, the investments required to prevent long-term project failure must be included in the stream of costs, or a concomitant reduction in the stream of benefits must be made to take into account a probable reduction in yields as the hazards advance.

CASE STUDIES AND MODERN PRACTICE

MANAGED AQUIFER RECHARGE

Mr Ian Gale from the British Geological Survey gave a realistic view of managed aquifer recharge stressing that often MAR is simply seen as a panacea for water supply problems worldwide and that not enough attention is given to the actual evaluation of the effectiveness of such structures. Thus, it was highlighted that MAR should be evaluated in its totality by considering its potential in different climatic, hydrogeological and socio-economic conditions.

This important evaluation of the impacts of MAR structures was depicted using 3 cases whereby 3 research sites in India, were established and monitored. These were Satlasana, Gujarat; the Kolwan Valley, in the Maharashtra state and Kodangipalayam in Tamil Nadu. All 3 sites were representative of different hydrological and socio-environmental characteristics and thus two studies; a physical and socio-economic one were carried out to obtain further understanding of each sites uniqueness.

Baseline surveys included geological mapping; the identification of hydrogeologically important features and topographical surveys. Several components such as rainfall, evaporation rates, groundwater levels and surface flows, climate, and sediment flow had to be monitored. In addition to this water samples had to be systematically collected to determine the quality of the water in the local wells in order to assess any possible change in water quality brought about by the MAR structures.

Attention was brought to the fact that the findings of the 3 sites are only preliminary since inter-annnual and annual variability in climate dictates that monitoring over a long-term period is required to obtain a more precise picture. From these initial findings, however, it was brought to the fore, that the effectiveness of each recharge structure in the states was dependent upon rainfall, geology and the actual usage of water abstracted.

In the Gujurat case, for instance, actual spilling of the check dam resulted due to the sudden and short duration of the torrential storm. Because of the coarse nature of the sediments in the headwaters of the catchment, rapid infiltration of water to replenish the aquifer resulted. Also, the stream flow was likely to have infiltrated further downstream if taken at a larger scale, thus reducing exposure time to be evaporated.

On the other hand at the Kolwan Valley site, the available storage was a constraint on the amount of additional water that can be recharged since the aquifer was of low storativity. Therefore recharge brought about by the dams has a short transit time since it resurfaces as baseflow. The highest increase in recharge was recorded in the Tamil Nadu site of Kodangipalayan since the aquifer is exploited to such an extent that natural recharge is insufficient to fully replenish the aquifer on an annual basis.

To conclude, MAR proves to be effective at replenishing over-exploited aquifers, and can also be used to control saline intrusion or land subsidence and improve water quality. However it was stressed that it only enhances the volume of groundwater abstracted and can play a role as part of a larger package when it comes to aquifer management, to control abstraction and restore groundwater.

ARTIFICIAL RECHARGE IN MAHARASHTRA, INDIA

Dr. Sunil Jain further contributed to the evaluation of managed aquifer recharge technology through his experience of artificial recharge in an overdeveloped watershed in Maharashtra, India. The watershed under study was the Bhaunak watershed which forms a part of the Tapi alluvial belt located within the basaltic Deccan Traps of Maharashtra State.

The multidisciplinary study integrated the science of water, land, topography, and climate, as highlighted previously by Mr. Ian Gale. Most of the area is cultivated (approximately 70%) and a rapid decline in groundwater levels (on average 1m decline per year) had sparked an interest in aquifer recharge possibilities. The over-abstraction of water is due to over 3000 dugwells and 500 tubewells that have resulted in an extensive overdraft and groundwater imbalance.

Various aspects of the study were explained in detail by Dr. Jain in order to reveal the significant role each study played in the final evaluation of the scheme. The morphometrical analysis of the watershed revealed that the value of the drainage density often indicates moderate to high permeability and relief in the watershed. The drainage length and slope are also required to define the number of water conservation structures that could potentially be installed in the mountainous area.

The role of geomorphology and geology of the area was also studied in significant depth, giving particular attention to the location of major faults and different strata composition within the basin. Climatological analysis was permitted through daily rainfall and rainfall intensity analysis, together with other weather components such as temperature, wind velocity, relative humidity and the number of sunshine hours. Soil texture and moisture was also studied in order to identify infiltration rates and thus designate areas ideal for the location of percolation tanks. Soil texture analysis is also deemed essential since this ultimately influences what managed aquifer recharge techniques should be chosen and the actual design of such techniques.

Continuous monitoring of percolation tanks and stream flows enabled the construction of a realistic estimate of the catchment yield in different catchments. It resulted that upstream mountainous stream flow was ideal for artificial recharge due to 2 main reasons - silt free clean water and flow available for a considerable period. Such recharge would augment both the quality and the quantity of water. In the lower catchment, higher silt loads were found, thus being inadequate for recharge due to in unreliability as a source and also because of the possibility of clogging any of the recharge structures with silt.

The construction of lithologs contributed towards the greater understanding of the hydrogeological framework and aquifer geometry of the site. It also permitted the realization that surface spreading techniques of MAR, such as percolation tanks, are not feasible in the part of the basin composed of granular and clay horizons. On the other hand MAR structures such as injection wells and recharge shafts were seen to be more appropriate.

Once the groundwater recharge potential was assessed and an impact assessment considering all types of MAR techniques, together with small scale rainwater harvesting through roof top harvesting, was carried out, a groundwater augmentation plan could be derived. Once again community participation was highlighted to be critical to ensure the success of such measures.

WATER QUALITY ISSUES AND RAINWATER HARVESTING

Prof Mike Edmunds elaborated on the significance of evaluating the quality of water when considering rainwater harvesting and its effectiveness. Rainwater harvesting can often provide good quality water in areas affected by water pollution and salinity and thus harvested water may provide an improved and safer supply. However quality problems may arise during the collection and actual storage of rainwater. Rainwater that is collected in a roof tank bypasses the natural mineralizing process and thus may be less acceptable for drinking purposes due to its acidity, low mineralization or due to contamination on storage.

The initial rain water chemical signature is derived from the source of water vapour, primarily the oceans, and contains solutes entrained as marine aerosols, atmospheric dust and any impacts of human activity such as forest fires or industrial emissions. This signature changes as soon as rain comes in contact with the ground. This is because of the uptake of solutes and nutrients. Rainwater is slightly acidic and low in dissolved solids, but on entering the soil it rapidly takes up carbon dioxide through microbiological processes. This helps in the dissolving of minerals. If carbonate minerals are present, the water will become close to neutral because of the buffering created by the carbonate-bicarbonate system.

Some published studies revealed that the quality of harvested rain in some developing countries does not meet the drinking-water guideline values since most water is contaminated microbiologically and thus special care has to be taken during the collection and storage of rainwater. Heavy metals and trace organics may also pose problems. In the case of groundwater recharge, most of the uptake of solutes in the water takes place in the unsaturated zone due to weathering reactions.

This process neutralises acidity and mineral dissolution can be enhanced as a result of the high soil pCO2 concentrations. As a result, the main properties of water quality are determined in the top few metres.

The quality of artificially recharged water is superimposed on the natural baseline water quality which may or may not differ from natural recharge. However in semi-arid regions baseline quality needs to be assessed since high quality, relatively modern water from artificial recharge may be superimposed on a baseline of older water containing, for example, higher salinity or high fluoride.

Environmental chemical and isotopic tracers are also often used to help demonstrate recharge pathways of recent rainwater recharge and also confirm the efficiency of managed recharge schemes. Prof. Edmunds mentioned that the most common and simple method was the chloride mass balance (CMB) method to estimate recharge. In this balance the increased salinity (chloride) is proportional to the amount of evapotranspiration and inversely proportional to the amount of recharge (R).

In many semi-arid regions this method has been applied successfully using unsaturated zone profiles, where surface runoff is near-zero. Including surface waters groundwater samples in the CMB approach is more complex but if the sources of Cl are known, an inventory of water inputs and outputs from the catchment or watershed may be involved and the overall efficiency of the recharge scheme should be related to an overall decrease in Cl. It is believed that there is scope for further application of the CMB to estimating the efficiency of rainwater harvesting schemes on condition that appropriate data are collected including that for establishing baseline conditions.

Stable isotopes of oxygen and hydrogen are also potential tracers in evaluating rainwater harvesting. Individual monsoon events may thus have distinct fingerprints that may be traceable into the aquifer. New storms may be distinguishable from previous season's rainfall which may have undergone evaporation and enrichment in the heavier oxygen isotope (180).

This approach needs testing on research catchments but should lead to better understanding of the recharge efficiency in a well characterised watershed, especially if tested with CMB and WTF methods. A case study on the quality of rainwater was presented for the state of Rajasthan, India, which is considered to be one of the worst-affected states in terms of high flouride contamination of aquifers. There is a high incidence of dental and skeletal fluorosis in the province.

The presented case study showed that in low-rainfall areas with potential fluoride problems such as Rajasthan, it is important to assess the hydrogeological situation very carefully. The generation of high-fluoride groundwaters usually requires considerable residence times in the aquifer.

Thus, it is likely that younger, shallow groundwaters, for example those recharged rapidly through check dams and stream channels, may have low fluoride concentrations compared with older groundwater. They may be exploitable by skimming the shallow water table rather than abstraction from deeper penetrating boreholes which mix water of different quality.

High fluoride waters have traditionally been treated by a range of techniques. However, village-scale fluoride removal presents drawbacks in terms of removal efficiency, cost, ethical issues, local availability of materials, chemistry of resultant treated water and disposal of treatment chemicals as well as monitoring the process. The harvesting of rainwater, either directly in cisterns, tanks or by careful collection via small recharge dams, offers a potentially safe and attractive alternative solution in endemic areas.

Professor Edmunds then summarized the main factors relating to quality of collected rain in relation to materials and construction, including roof materials, roof location, avoidance of matter deposited on roofs and avoiding anaerobic decomposition of organic matter.

The chemical and microbiological quality of collected water was then considered and how best to avoid pathogens, trace organics, mosqitos and other contaminants

COSTS AND BENEFITS OF RAINWATER HARVESTING IN IRAN

Further evaluation of the potential and impacts of water harvesting were given by Dr. Sharifi through his recount of experiences and lessons learnt from rainwater harvesting in Iran.

The full account of the cost-benefit analysis described by Dr. Sharifi enabled the participants to clearly see what the potentials and limitations of RWH are in a semi-arid and arid environment, as revealed through Iran. Dr. Sharifi described rainwater harvesting as an inexpensive means of mitigating water supply deficiencies.

The cost-benefit analysis carried out revealed that RWH is an economic and efficient method since it results in a large return for relatively small investments. What makes its so attractive economically is its simplicity in technology and thus its applicability that is added impetus in encouraging the community to actively participate in RWH promotion and implementation.

When exposing the vast potential Iran has in terms of water resources, Dr. Sharifi gave estimates from a surface flow analysis, a 30-year rainfall analysis and a groundwater balance analysis. All three studies revealed that the potential water from these water resources was highly significant. For surface flow an estimated additional 260Mm^3 was calculated.

The potential yield of water resources in 646 monitored sub-catchments in wet years was higher than 228 Mm^3; whilst the groundwater balance study taken across 3 watersheds in Iran, resulted in an average annual availability of 5 Bm^3.

In view of this, the scarcity in water resources in Iran can be attributed to watershed and aquifer mismanagement. It was pointed out that management of these resources could only lead to important soil and water conservation, which in turn will yield a cumulative positive effect on the maintenance of soil nutrients and organic content, as well as the propagation of plants.

Dr. Sharifi also mentioned the utilization of floodwater as an important solution to the shortage of fresh agricultural water. It was argued that this floodwater can best be stored in mountainous areas due to the reduced evaporation losses in such environments. Flood spreading over coarse-grained alluvial fans also enhances agricultural production and forest reclamation.

Over 50% of Iran is mountainous, with very low evaporation capacities and also covered in coarse-grained areas, thus providing the capability of storing water and delaying sub-surface flows. It was stated that if 5 rainfall events occur annually and only 1000m^3 of water is infiltrated, then the total precipitation in a wet year can be completely accumulated in natural reservoirs. It was also argued that investing in new dams does not make sense since evaporation rates in open reservoirs in high and dams face problems of sedimentation and pollution, thus increasing costs of maintenance.

Also the investment required for each cubic meter of water storage capacity in surface open reservoirs is nowadays 5 to 10 times higher than the capital investment formerly required for similar projects constructed in the past. Moreover in the face of the uncertainties brought about by climate change the estimated storage capacity of these dams cannot be justified due to fluctuating annual precipitation.

WATER HARVESTING TECHNIQUES IN THE ARAB REGION

Dr Zakri gave an overview of the diversity in water harvesting techniques in the Arab region, explaining the use of structures in Yemen, Sudan, Libya, Tunisia and Egypt. The main point exposed through Dr Zakri's presentation was that water harvesting is central to alleviating water scarcity problems in the Arab region. This explains why diverse water harvesting techniques have been widely practiced in the Arab region dating back over 9000 years.

The catchment area in Arab countries is a major criterion for classifying water harvesting systems as follows:

- Micro-catchment water harvesting systems where the catchment area and cultivated area are adjacent. Eg. The Negarim microcatchments and contour bunds are examples of this water harvesting technique.
- Macro-catchment water harvesting systems where the catchment area is located upstream of the cultivated area, permitting the harvesting of overland flow.
- Spate irrigation system which depends on harvesting flood water from wadi channels. This type of RWH technique requires a huge catchment.

Dr. Zakri explained a wide range of RWH systems in the Arab world, some of which were further described by Dr. Noman in the subsequent presentation.

A full account of the terraced systems in Yemen and spate irrigation systems in Sudan was given, exposing the ingenious technical and engineeing capabilities of these seemingly simple but effective techniques and structures. Of particular interest was the description of the Miskat systems, (or *Meskat/Jessours* in Tunisia), one of the most ancient RWH methods employed. The use of these systems is common in the Arab Maghreb (Tunisia, Morocco and the north-west of Libya in Jebel Nafousa).

The Miskat was described to be simply a piece of flat land with a mild slope (3 to 6%) with few or no drainage channels. The land is prepared for rain water harvesting and then water is directed to another piece of land of half its area and located directly below. This parcel of land is called the collector where crops are planted. Water can thus be said, to be stored in the soil horizons. Unfortunately the state of these Miskats has deteriorated because of the intensive agricultural development that took place since the middle of the century.

Two case studies from Egypt were also presented by Dr. Zakri. The first looked at Wadi Watier in south-eastern Sinai. This wadi is characterised by strong flash floods that have put several lives at stake and destroyed several areas which are of historic significance. For this reason 17 retention dams and 5 storage dams have been proposed. The costs of these dams have been calculated to be compensated within a few years after the execution of the proposed control dams. The second case described the Wadi Ghuweiba, in the Eastern Desert in Egypt.

This wadi consists of Eocene limestone outcrops forming a delta. The delta itself is characterized by a gradually sloping irregular surface dissected by fan drainage lines and covered by alluvial deposits, considered as an important source of groundwater in Quaternary sediments that can be withdrawnl by shallow dug wells taking into consideration the sea water intrusion from the Gulf of Suez due to over pumping.

Geoelectrical resistivity sounding was employed to determine the thickness of the layers. From this study it resulted that there two main aquifers in Wadi Ghuweiba area; the upper Quaternary aquifer which can be harvested by drilling wells to a depth of about +150 m, and the Tertiary aquifer which can be harvested by drilling wells of about +450 m. In order to increase the rate of recharge to the Quaternary aquifer, an artificial recharge system was recommended.

Six locations were employed to induce infiltration into the Quaternary aquifer using of instream structures. These structures were series of rechargeable dams which are changing the hydraulic regime of wadi Ghuweiba stream, decreasing flow velocities and encouraging the growth of riparian vegetation..

Dr. Abdulla Noman from the Water and Environment Centre (WEC) of Sana'a University, Yemen gave a comprehensive overview of the various state of the art techniques used in Yemen for using and managing harvested water. Water collecting catchments are often located away from village cores in order to prevent pollution.

Yemen, having historically been one of the oldest civilisations in the Middle East that depended upon agricultural development and thus the main purpose of use of harvested water in cistern structures was for animal watering, drinking and supplementary irrigation. This has permitted Yemen to be self-sufficient in food production for centuries and moreover export its surplus. Changes however, in the lifestyle and consumption patterns of the increasing population has caused the abandonment of terraces and their maintenance and the absence of traditional cooperation among farmers due to both internal and external migration.

Dr. Noman mentioned the production of a local material known as *khadad*, which was and is still used to cement rainwater cisterns.

This material has proven to be of high quality due to its resistance against weather elements and the guarantee of its long lasting quality. Further research on the application and production of this material is required.

The variety of rainwater harvesting systems in Yemen include roof top harvesting whereby water is collected in a dug out structure called a *Seqaya.* In hilly areas rain is collected on the roof and directed into an underground storage tank through a settling tank used for domestic use.

Terracing is also utilised in the mountainous areas whereby terraced fields provide sufficient time for rainwater to intercept the soil, and low-lying contour built rubble walls prevent runoff but simultaneously permit the water to move down slowly from one terrace to the next through the voids between the rocks in the wall.

In addition to terracing, ponds for agricultural use are also used. These can be either excavated farm ponds, embankment ponds or excavated-cum-embankment ponds.

Cisterns are also used particularly in the mountainous areas of Yemen. These cisterns are usually underground masonry tanks used for drinking and domestic purposes.

Flood water harvesting and spate irrigation is common practice in Yemen.

This technique involves the directing of floodwaters in mountainous areas to coastal and foothill areas. In this way spate irrigation is utilised for the production of major crops. Such a traditionally embedded technique is characterised by a water right given to upstream users over downstream users, such that upstream irrigator's only releases water to downstream users once all the water needs have been satisfied.

The irrigator is then obliged to release water to downstream irrigators.

Diversion structures used in spate systems are needed to direct large flood waters away from the command area so as not to cause damage.

The structures utilised are deflectors or *Al-Qaid* (low earthen bunds); high earthen bunds (*Ogma*); drop structures (*Al Masaqit*) and spillways (*Al Masakhil*).

Water rights are not consistent in Yemen. This implies administrative difficulty when it comes to keeping track of legitimate users. However a person in each area (known as an Al-Moqadem), a local expert, would know the area and land owners very well and thus aid in the understanding of traditional law of water rights. Dr. Noman ended his presentation by giving his views on the future role of rainwater harvesting.

From the Yemen experience appropriate systems can evolve form traditional know how on water harvesting. It was again emphasised that lessons should be learnt from the shortcomings of previous projects.

Worldwide the development in rainwater harvesting technologies has aided the scientific and management community to see such technologies as supplemental water systems and combined system that allow the prolongation of the cropping season or even the possibility of growing a second crop.

Rainwater harvesting is also seen to have a dual purpose since the first runoff waters can be used to irrigate cropping areas and excess water can then be stored in some facility for later irrigation use.

Rainwater harvesting potential to enhance soil water storage is also essential since the water retention capacity of soils has to be high enough to supply crops with sufficient water, especially in dry environments when rainfall is intermittent. Lastly the importance of a further need for in depth knowledge on the hydrological, pedological and crop parameters is required so that more efficient models could be developed.

WATER HARVESTING EXPERIENCES IN THE INDIAN SUBCONTINENT

The final group of presentations all struck the same chord—the significance of NGOs in the role they play to encourage rainwater harvesting at the community level.

Ms. Kavarana (CSE Delhi) and Mr. Sharma (Wells for India), both representing highly influential organisations gave insight on how local communities can actually be made to feel a part of the implementation of RWH in their own environments. A case study of the Aravali Hills Rajasthan India was presented by Mr. Sharma. Wells for India is a UK registered Charity that works with poor communities in Rajasthan, India.

The charity supports RWH as a primary intervention in community development. The case studies illustrated that safe potable water availability provided by rainwater harvesting schemes not only improve health and releases time and energy of women, but also improves poor people's income, education, as well as social and cultural well-being.

Increased water availability through small scale water harvesting structures in *Nayagaon, Jamun and Nala* villages has increased the productivity of crops, fodder and milk production. Local village self-help groups also started playing active roles in the village development issues such as education and health care. The communities are carrying out many activities themselves, enabling to take next step towards better sustainable future.

This work is spread over 11 districts of Rajasthan working with with 22 local non-government organizations (NGOs). Projects are generally based in village clusters in the upland areas of a river basin where resources can be targeted at individual villages.

Successful water harvesting intervention in one village often provides the model for adoption and replication in neighbouringneighboring villages. Results from a 5-year water harvesting project show that RWH is an essential foundation for all other forms of development in a village.

Small scale water harvesting work not only helps in increasing water availability but also in enhancing productivity of food grain and fodder and allowing income generation.

Local village self-help groups formed in the villages are playing active roles in the village development tackling issues such as health and education. Through water security groups are motivated to carry out a variety of activities, enabling them to take their next step towards better sustainable future.

Rainwater Harvesting in the Cholistan Desert, Pakistan

Dr. Mohammad Kahlown of the Pakistan Council of Research in Water Resources gave an interesting overview of the Pakistani experience on rainwater harvesting initiatives that were created in the Cholistan desert, Pakistan.

About 70 million hectares of Pakistan fall under arid and semi-arid climate including desert land. Cholistan is one of the main deserts covering an area of 2.6 million hectares where water scarcity is the fundamental problem for human and livestock populations as most of the groundwater is highly saline.

Rainfall is the only freshwater source, which occurs mostly during the monsoon (July to September). Therefore, rainwater harvesting in the desert has crucial importance.

The Pakistan Council of Research in Water Resources (PCRWR) has been conducting research studies on rainwater harvesting since 1989 in the Cholistan desert by developing catchments through various techniques and constructing ponds with different storage capacities ranging between 3000 and 15000 m^3.

These ponds have been designed to collect maximum rainwater

within the shortest possible time and to minimize seepage and evaporation losses. As a result of successful field research on rainwater harvesting system, PCRWR has developed 92 rainwater harvesting systems on a pilot scale in Cholistan desert. Each system consists of a storage reservoir, energy dissipater, silting basin, lined channel, and network of ditches in the watershed.

The storage pond is designed to collect about 15000 m^3 of water with a depth of 6 m. Polyethylene sheets (0.127 mm) on the bed and plastering of mortar (3.81 cm) on the sides of the pond was provided to minimize seepage losses.

All these pilot activities to harvest rain have brought revolution in the socio-economic uplift of the community. These activities have also saved millions of rupees during the recent drought. Large scale adoption of all these interventions would ultimately help improve the socio-economic conditions of the residents of hyper-arid area of the country.

Chapter 7

Integrated Health and Agriculture Extension in Dry-Land India

Extension and education are critical components in development across a wide range of issues, including natural resource management, agricultural productivity, health and sanitation, and watershed management. In order for rural development to be successful, it must be approached from a holistic point of view, and each one of these issues represents an essential part of integrated rural development. Multifunctional extension programs are useful strategies for accomplishing integrated rural development, and can include environmental, economic, health, and agricultural productivity goals.

Our chapter is designed to discuss how Indian extension can better accommodate a variety of rural development strategies with agriculture at the core. Specifically we discuss agroforestry, nutrient management in soil conservation, watershed management, and extension programs focusing on linking agriculture and health. In all of our discussions there is an implicit understanding that the technical aspects of extension and training must be married to the socio-economic characteristics of communities where extension agents are working.

For some of our papers this understanding is an essential facet of our discussions and the subject of our analysis and critique of extension programs in India. Much of our analysis focuses on India's dry-land regions, where low levels of agricultural production and high levels of environmental degradation are inextricably tired to rampant poverty and malnutrition. Thus, an integrated rural development package must be adapted to local conditions in order to effectively address the multiple and interlocking barriers currently facing the rural populations living in these regions.

IMPROVEMENTS IN AGROFORESTRY EXTENSION IN INDIA

Agroforestry has the potential to significantly improve the livelihoods, economic viability, and agricultural production of small farmers in traditionally marginalized areas. In addition, agroforestry can be used as a tool for changing attitudes towards natural resource management.

In order to understand how extension can better accommodate natural resource management through the successful introduction of agroforestry and tree planting, extension agents, both public and private, must strive to understand the socio-economics of agroforestry and of the communities within which they work ("On farm..."). While agroforestry has traditionally been successful in humid and semi-humid areas, it is most needed in dry-land areas, which are

characterized by acute climactic challenges and extremely low levels of agricultural production ("Agroforestry Systems in India" 2004 and "On farm..."). The limits to the adoption of agroforestry are many, but must be viewed beyond technical and biotic constraints of the dry-land agricultural regions; the ability of extension agents to recognize socio-economic characteristics of farmers is essential in learning how to target potential innovators who can then generate change within their communities.

More emphasis on the development of participatory programs, more efficient targeting, and intensified focus on the traditional communication channels will increase the adoptability of agroforestry in dry-land communities. In turn, an implicit understanding that the attitudes of farmers towards natural resource management are based on agroforestry successes and failures will help extension to increase the environmental reclamation of degraded landscapes, especially those in dry, rain-fed agro-ecosystems.

Dry-land agricultural areas in India face specific problems that must be overcome for both improved agricultural production and environmental conservation. Agroforestry can help overcome limited water availability, fuel shortages, degraded soils, and a lack of biodiversity ("Agroforestry Systems in India" 2004 and CRIDA). When trees are used that are adapted to water-limited areas, such as Gliricidia, communities can realise both agricultural and environmental benefits (CRIDA).

In general agroforestry produces a multitude of benefits, including increased income from selling surplus goods, improved environments and enhanced biodiversity, increased access to education through income generation, shelter from harvesting trees for building materials, improved health when medicinal plants are incorporated, a greater diversity of food sources, and an increase in the energy of an agricultural system because manure can be used for fuel instead of food ("Agroforestry Systems in India" 2004).

Specifically in rain fed agro-ecosystems, like those found in dry-land agricultural areas of southern India, successful agroforestry projects can increase the organic matter in soil, improve the physical properties, increase crop yields, increase the availability of nitrogen in the soil, increase the water-holding capacity of soil, and reduce soil erosion. In addition, there is a reduction in pollution because of a decrease in chemical fertilizers.

Most environmentally significant, the biodiversity of an agricultural area is enhanced (CRIDA). Agroforestry's scientific benefits as listed above have been well documented within the humid and semi-humid agricultural regions, and the challenge for extension agents is to successfully spread these benefits in dry-land areas where, traditionally, farmers have been marginalized due to resource limitations (such as soil-fertility and water availability). These benefits are of utmost importance in creating a sustainable agro-ecosystem in dry-land agricultural areas, and they cannot be realized without appropriate extension and education.

Most importantly such extension and education must be localized and participatory in nature (Glendinning et al. 2001 and "On farm..."). In order for training and education to be localized, extension agents must understand the important role of socio-economic factors as well as the technical aspects of agroforestry, and therefore they must begin to focus more on the social and communication constraints of the specific villages and communities within which they work. Indeed the foundation of success is the "participation of the people for whom the forestry programme is intended".

This participation of local residents will help transform the extension programs from the current "top down" structure, pioneered in "transfer of technology" methodology, to "bottom up" participant defined priorities and goals. In this way, participants are much more responsive to the education and training offered by extension agents. In fact, the potential participatory nature of agroforestry programs in rural communities can be a useful vector for knowledge generation that incorporates local information and global science.

Villagers and communities, depending on a number of variables, are continually gathering knowledge from various sources, and extension agents should do less to emphasize the strict dichotomy of global science and indigenous knowledge as it can be harmful to the adoption rates of tree management and agroforestry.

Such a dichotomy will ultimately only deepen the impression of 'otherness' of indigenous and rural people's knowledge and thereby reinforce the developed/underdeveloped polarization that many of these same researchers and development experts are trying to overcome...Rather than solely emphasizing differences, scholars might accomplish more by seeking out points of congruence and interface between local realities and larger-level systems.

Ultimately, Brodt and others advocate that synergy instead of trade-offs be emphasized when extension agents use a variety of knowledge sources to advocate certain practices. Communities within dry-land regions of India have already developed knowledge concerning tree management; however, these knowledge systems are "open", thereby allowing an exchange of information which can be more conducive to innovation than enhancing a "closed system" through strict juxtaposition of local knowledge with global science.

By allowing the participants to innovate beyond their previous knowledge capacities, extension agents are facilitating local ownership of knowledge, which enhances the success of programs and promotes long term commitment on the part of the farmer and community. Participant defined priorities and goals and an incorporation of local knowledge by extension agents must also be complimented by designing programs that better target potential users, early adopters, and innovators.

Programs must be designed to address different family units, income levels, ages, land tenancy, farming experience, castes, occupations, urban contact, degrees of social participation, farm modernity, household assets, access to capital, and varying off farm incomes. Methods have been developed that, when using these and other variables, can successfully identify those farmers that will be "tree planters" within a subsistence farming community. It is essential that, before an extension programme is implemented, "the beneficial characteristics that might promote or impede local involvement in forestry programmes" are defined.

Mahapatra and Mitchell have identified several key factors that make a farmer into a potential "tree planter" including "progressive attitude", "membership in village organizations", "wealth status", and "perceived risk concerning agricultural production." In addition to such characteristics, they also note that off farm income generation is generally very complimentary to the low labour inputs of agroforestry (or home gardens). Most importantly, it has been shown that social or economic factors cannot account for the adoptability of tree planting and agroforestry; these factors must be combined in order to successfully target farmers who will plant trees.

Another important note is that land area and tenancy are important, but mistakenly overemphasized when targeting farmers:

"Classification of farmers by size of landholding, as [is] usually done by agricultural extension workers in India, is not suitable for farm forestry extension". This is especially important for agroforestry extension in dry-land areas where farmers are operating on very small plots of land—further supporting the spread of agroforestry beyond humid and semi-humid areas.

As discussed before, an exchange of information and knowledge generation are both critical in innovation and the success of agroforestry, and the model developed by Mahapatra and Mitchell also emphasizes this critical point by observing that membership in village organizations and contact with urban populations are essential indicators in targeting adopters and innovators.

By better understanding how to target specific farm types, farmers, and communities extension programs can shift away from the "conventional approach" of delivering standardized tree seedlings and vague management advice to all farmers.

In this way they can concentrate their energy, time, and resources on specific farm types and farmers as well as identify the "uncertain" group that could either adopt or dismiss agroforestry. By helping change the attitudes of "fence riders" and targeting the farmers that are very willing to innovate, agroforestry programs will see much more long term success and increase the adoptability of tree planting and, thus, natural resource conservation.

The participatory model of extension and education advocated throughout this chapter as well as the emphasis on targeting potential adopters and innovators is underscored by a better understanding of local knowledge dissemination and technology transfer among rural farmers in India's dry –land regions. When extension agents invest time in learning how villages and farmers communicate, they also see that participatory programs and targeting initiatives are critical in establishing successful agroforestry farms; "the decision to adopt agroforestry was found to be determined by the farmers' attitude to agroforestry, which in turn was shaped by information received through farmer-to-farmer and farmer-to-extension contact.

The mode of communication was important and, to be effective, needs to be customized for each target group". Glendinning has found that the most important channels of communication for farmers, when innovating and learning, are extension agent, neighboring farmers, and group meetings.

These channels of communication indicate that participatory programs and better targeting by extension agents are appropriate ways to disseminate knowledge and change attitudes. Agents can target potential innovators who are capable of bearing more risk than other farmers. When these innovators generate both economic and environmental benefits through agroforestry, it is more easily communicated to neighboring farmers and members of group organizations.

Neighboring farmers are more likely to adopt agroforestry if they see it has been successfully done by members of their own community. Successful demonstrations by villagers and other farmers decrease the perceived economic risk of agroforestry. "In regions where smallholders predominate [like dry-land areas in India], promoting change within the existing farming system may therefore require substantial diversion of resources in order to develop effective communication systems, skills, and media".

It is therefore equally important for extension agents to assume the role of facilitators when first entering communities; they can create or enhance pre-existing village meetings and farmer-to-farmer contact. Extension agents can take advantage of the local social structures to increase the "adoptability" of agroforestry.

In addition for small landholders personal communication with neighbors and extension agents are the two modes of communication and knowledge sharing that are most strongly linked to potential adoption and tree planting. Understanding the communication channels and constraints will enable extension agents to enhance village communication and capacity building of farmers, which will, in turn, help demonstrate the low economic risk of agro forestry as well as the environmental benefits and improvements in agricultural yields.

Local knowledge, participation, and better targeting by extension agents are critical in establishing a long term commitment to agroforestry in dry-land regions of India. The importance of successful agroforestry extends beyond satisfying agricultural production and economic goals, it is also critical in establishing an understanding of why natural resource management and environmental conservation are important in the sustained productivity of ecosystems.

The future of natural resource management depends on the attitudes of farmers which are distinctly shaped by the information they receive from communication channels guided by extension services.

Extension agents in India have established the scientific base and technical background to support agroforestry adoption; however they have yet to master the socio-economic tools to analyse communities and enhance local social structures that enable the adoption of agroforestry and facilitate knowledge exchange and generation, critical components of successful agro-ecosystems.

If extension agents can use socio-economic analysis to better understand the local conditions in which they operate, the adoption rates of agroforestry will increase, and the overall environmental benefits will be realized.

These environmental benefits are critical to sustaining agricultural production to feed a burgeoning population. One of these, nutrient management in soil fertility, is of overwhelming importance for cropping systems. Further examination of nutrient management in Indian dry-land agriculture is needed to understand the future of agricultural productivity, and the overwhelming importance of agroforestry in degraded landscapes.

Such novel approaches aimed at better understanding local conditions and farmer capacities and attitudes toward adopting improved land management are also a useful tool for analyzing nutrient management issues in much of India's degraded dry-land agricultural soils.

As we will see in the next section of our group paper, agro-ecological constraints to enhanced food production often are exacerbated by poor land management practices among resource poor farmers living in these areas. Thus, new soil and cropping systems management must be adapted to local conditions in order to ensure sustained agricultural production without depleting the existing natural resource base.

SOIL FERTILITY MANAGEMENT IN DRY-LAND INDIA

Expensive soil and water management technologies, high capital and energy requirements and "top-down" development approaches have not alleviated soil fertility problems in Indian dry-land agriculture. Rather, in spite of many costly investments in agricultural extension, soil degradation worsens and production remains extremely low. On the technology side, the majority of marginal and small-scale farmers throughout the region are ill equipped to benefit from new farming technologies that often promote the excessive use of agrochemicals and other external farm inputs.

Thus, future thrusts in agricultural extension should focus on promoting more sustainable alternatives to India's ongoing downward spiral of declining soil fertility and low crop yields. One effective tool for analyzing such approaches on both a theoretical and practical level is to learn from the vast body of available indigenous knowledge on soil fertility management. Traditional knowledge and practices related to dry-land agricultural production have been adopted and modified by Indian farmers based on local needs and resource availability.

Advances in dry-land agricultural management must seek to build upon- rather than replace- indigenous soil fertility management practices in a cost effective and sustainable manner that allows for both enhanced agricultural production levels and improved natural resource management. Most of India's dry-land regions consist of two primary soil orders- Vertisols and Alfisols- both of which pose significant problems for agricultural management.

Vertisols, with their characteristic shrink-swell behaviour of clays as they go through wetting and drying cycles, have low structural stability and thus, are difficult to manage. Due to the critical timing of performing tillage operations on these soils, some farmers have dubbed them "24-hour soils" for their tendency to be too hard and dry to cultivate one day and too wet and sticky the next (FAO website, accessed 3/05).

Alfisols, meanwhile, are characterized by very low natural levels of soil organic matter (SOM) and nutrient availability. They also have poor water holding capacity and are prone to extremely high rates of surface runoff and erosion when left bare under heavy rainfall conditions, such as those seen during India's monsoon season. Thus, agricultural production on these soils is extremely low and land degradation very high.

Traditionally, agricultural extension in dry-land India has focused on improving overall production levels by eliminating or reducing the effects of a number of agroecological factors. In dry-land areas in India and elsewhere, lack of adequate water supplies and poor water management are the most prominent limiting factors for improving agricultural productivity. Public extension in India's dry-land have thusly invested many resources into high input irrigation systems and rainfall collecting and harvesting technologies.

Comparable investments have been made to promote and disseminate new high yielding seed varieties (HYVs) of basic grain crops and cereals, such as rice, wheat and maize. HYVs have fared very well under favorable on-farm conditions with the application of external inputs, but have not been as effective for resource poor farmers in India's dry-land regions. Much of this is due to the lack of available land, labour and capital needed for purchasing and effectively managing the technologies.

Poor land management practices also have significantly contributed to the concomitant problems of loss of soil fertility and low agricultural yields. Deforestation and overgrazing are the two primary factors contributing to soil degradation, and crop yields are further limited by mismanaging tillage operations and fertilizer applications as well as mistiming planting dates for optimal harvest. Research has shown that up to 40 tonnes per hectare of valuable topsoil may be lost following 15 years of cultivation. Much of this loss may be attributed to intensive tillage operations that leave the soil bare and unprotected for long periods of time.

As soil organic matter (SOM) is washed away by wind and water, soil fertility is reduced and crop yields subsequently decline. SOM content can be replenished and greatly enhanced by simply not removing crop residues and farmyard manure (FYM) from the system; they are left to decompose in situ or incorporated into the soil profile. However, high demands for fuel and animal fodder result in the removal of many valuable nutrients from the agricultural system, and soil fertility remains low.

In spite of some of traditional dry-land agricultural systems' shortcomings, there exists a vast and growing body of knowledge on indigenous soil fertility management practices. Farmers in Karala and Karnataka have been planting vetiver grass (Vetiveria zizanoides) for more than a century as a means of erosion control, and farmers throughout various dry-land regions are reported to have been incorporating animal and green manures into the soil and practicing agroforestry and intercropping.

Research shows that such practices are most effective when combined with the judicious use of crop rotations to suppress the prevalence of weed, pest and pathogen problems. Some of these indigenous practices, such as the planting of Prosophis juliflora in alkaline soils, have proven highly effective in the reclamation of low fertility, degraded soils throughout India's dry-land regions.

If Indian dry-land agricultural extension hopes to increase productivity levels without further depleting soil fertility, scientists and extensionists must better understand the important role of indigenous knowledge and soil fertility management practices in achieving such goals. Local "best-practice" technologies have the inherent advantage of being adapted to local agroecological and socio-economic conditions.

In contrast, top-down technology transfer (ToT) approaches are insensitive to such important dry-land agricultural variables as rainfall, soil type, vegetative cover, available farm inputs and off-farm nutrient demands, and farmer adoption rates of new technologies remain alarmingly low as a result. Thus, extension must look to complement- rather than displace- indigenous knowledge and local best practice in

India's dry-land regions. Doing so increases the likelihood of new technologies being adopted and modified by farmers, and promotes managing local resources more sustainably and productively.

Researchers and extensionists have been experimenting with more participatory approaches in dry-land agricultural production. This change has coincided with increasing farmer use of private, fee-based extension systems in India. Participatory rural appraisal (PRA) is being used to collect, synthesize and analyse available information on indigenous soil fertility management.

This information will be used by extensionists and policymakers to promote resource-efficient agricultural development that is more compatible with the agroecological and socio-economic conditions faced by both large- and small-scale farmers alike. Private, fee-based extension programs have the advantage of direct accountability to the farmers they serve, but lack of money to pay for their services is a problem for many small-scale and marginal farmers.

Dry-land agricultural extension, whether public or private, must incorporate scientific knowledge into indigenous knowledge systems and local best practices in order to meet the needs of both better endowed and resource poor farmers throughout the region.

One example of a current agricultural extension initiative that is attempting to incorporate both scientific and local knowledge and practices for managing soil fertility is the current promotion of conservation agriculture (CA). Rice-wheat farmers in the Indo-Gangetic Plains have successfully adopted and modified the no-till package for wheat cultivation, though many farmers continue to use traditional tillage and flooding practices for cultivating rice.

CA advocates the permanent use of no-till technologies and practices while also maintaining permanent or semi-permanent soil cover and using regular crop rotations. Not surprisingly, researchers and extensionists are finding no-till adoption most successful in regions already having a strong tradition of growing green manure cover crops (GMCCs) and adding farmyard manure to enhance soil fertility. These factors are strongly influenced by several socio-economic factors such as high demands to feed crop residues to animals and use manure for fuel.

History demonstrates that new technologies relying on high capital and energy requirements are unlikely to significantly impact resource-poor farmers and their families. Examples of traditional soil fertility management in India portray a wide variety of local best practices, many of which are capable of increasing on-farm productivity without depleting or otherwise damaging natural resource bases. Similarly, public and private agricultural extension systems must seek to promote an effective balance between improving on-farm productivity and sustainable resource management. Furthermore, the technologies must be available and practical for farmers of all sizes and resource allowances.

In some dry-land regions of India, conservation agriculture practices such as the use of no-till, green manure cover crops and regular crop rotations have been successfully adopted by both large- and small-scale farmers to help alleviate soil fertility and production problems. Indian farmers that practice no-till and other components of conservation agriculture have reported decreased water, fertilizer and labour requirements and increased levels of production.

The resulting increased profitability for Indian dry-land farmers is the primary cause for widespread adoption. By increasing agricultural productivity and profitability while also enhancing soil fertility, conservation agriculture practices have tremendous potential to reverse the downward spiral of declining soil fertility and low crop yields. They are also highly capable of being adapted to local needs and levels of resource availability.

Some of the technologies also require enhanced levels of knowledge and training, and extension services would be responsible for ensuring that new information and materials are made locally available and practical for dry-land farmers. Researchers and extensionists must also seek to improve no-till rice production, and introduce new cover crop species that produce enough biomass to provide for soil cover and animal fodder in dry-land silvopastoral systems.

Such advances are likely to be adopted most readily when researchers and extensionists use participatory methods to involve local farmers and build upon preexisting indigenous knowledge systems and practices. Yet for soil fertility management to be effective in the long run, it must be part of an integrated community natural resource management programme that includes laws prohibiting the clearing of forests and the burning of crop residues as well as a plan for community watershed management. Limited water supplies and poor water management are the two primary limitations to agricultural productivity in India's dry-land soils.

Chapter 8

Watershed Management for Dry Land Agriculture in India

During the past five decades of planned development, India has achieved spectacular increase in food grain production. Green revaluation enabled us to enter in to a new era in input use. The use of high yielding variety seeds, fertilizers, plant protection chemicals have stepped up manifold. Development of major and minor irrigation projects had gone up. But still the over all gains had not been evitable shared by the farming community.

This is mainly because two thirds of agriculture in our country ifs rainfed. According to water for irrigation would emerge as a critical bottleneck. The requirements of groundwater, it will lead to over exploitation of ground water in several status and by 2020 as much as 71% of irrigation area is likely to be from ground water sources. Closely related to this is the problem of degradation of land in the rainfed areas.

Rain fed agriculture in India is characterized by low productivity, degraded natural resources and wide spread poverty. Most of millions of people living in our country depend on agriculture and natural resource management for their livelihoods, so development planners are eager to implement productive, environmentally sustainable land and water management. It is in this concept of watershed development has been introduced in our country.

WATERSHED

It could be defined as an area from which the runoff drains through particular point in the drainage system. Main aims of the watershed management programme in rainfed areas are:

- Enhancing crop production and increasing farmers' income.
- Minimizing degradation of natural resource base.
- Principles of watershed management:
- Utilizing the land according to its capability.
- Adequate vegetative cover during the rainy season.

- Conserving as much rainwater as at the place where it falls.
- Effective utilization of surface and ground water resources.
- Adopting integrated soil, water, nutrient, and pest management options.
- Increasing cropping intensity and land equivalent ratio through intercropping and sequential cropping.
- Safe utilization of marginal lands through alternative land use pattern.

Water Management Structures:

- Waterways
  - Waterways are made between fields to safely dispose off excess rainwater out of the field without much soil erosion.
- Farm ponds
  - Grassed waterways lead to farm ponds where excess rainwater is collected for supplemental irrigation during the dry periods or for animal use.
- Check-dams and gully plugging
  - Check-dams and gully plugging control soil erosion as well as facilitate percolation of rainwater into the ground.
- Percolation tanks

Percolation tanks such as check –dams and gully plugging help in reaching the groundwater. In our tour programme we were visited Kothapalli village in Andrapredish.it is organized by the international crop research institute for the semi-arid tropics ((ICRISAT) has launched a scheme on watershed management.

Which offers a pivotal linkage between the farmers needs resources and local environment and an institute. Under this condition ICRISAT identified Kothapalli village and established watershed at 2000. The over all aim is to combine the generalized information to suit profitably to individual farming situations. Different department people joined together and maintain the watershed in Kothapalli village.

Rain fall and Soil type:

The climatic condition of this village is hot, semiarid region with average annual rainfall is 500-600mm.Most of the annual rainfall is received during the June-Oct. 80% of rainfall is received during the summer season.

Off-season rain is used for preparation of land. In Kothapalli village the soils are widely distributed. The major soil types are black soil and incept sols. Maximum rainwater loss through evapotranspiration and runoff because the very low moisture holding capacity of the soil.

Cropping pattern:

In Kothapalli village-cropping pattern has changed, before the establishment of watershed they were cultivated only one crop, now they are cultivating three crops.

- First crop- Paddy.
- Second crop- Sorghum, Safflower, Sunflower, Pigeon pea.
- Third crop- Vegetables.
- It shows management the effective utilization of land.
- Benefits of proper watershed management:
- Increased crop productivity and profits.
- Reduced soil erosion and better land quality
- Improved surface drainage and reduced water logging.
- Increased ground water recharge
- Increased water availability for drinking for humans and animals.
- Long-term sustainability of agriculture.
- Improved agro-ecology of watersheds.

I concluded from that study is the watershed management is increased the water availability for crop cultivation. It enriched the ground water source.

The cropping pattern is changed and it has increased the standard of living of farmers. So, watershed management is one of the best schemes to improve the socio-economic status of Kothapalli village in Andrapradesh.

HUMAN DEVELOPMENT THROUGH INTEGRATED HEALTH AND AGRICULTURE

The previous sections have dealt explicitly with agricultural management systems in India and demonstrated the importance of considering multi-strategy solutions in agriculture development. The severe environmental constraints in dry land agriculture merit a dynamic approach to agriculture improvement that encompasses not only nutrient management and agro-forestry, but particular attention to local watersheds.

Though governmental extension services in India have adopted agroforestry and some watershed management programs in their extension training, little emphasis has been placed on human and environmental development where decisions and action plans are decided at the community level.

Public extension in India does not have the resources to implement integrated programs that value the participants' development even more than the farm systems' development. In response, private extension, primarily in the form of non-governmental organizations (NGOs), has begun to expand beyond the reaches of the production-oriented public sector, both in regional coverage and methodology. Integrated conservation and population programs are based on the principle that human health and development is essential to the sustainability and the protection of ecosystems, natural resources, and communities.

An analysis of the global relevance, the principles of human and integrated development, their advantages, and a case study of community based population and environment programme elucidates the magnitude of change integrated programs can generate. By nature, communities are the interface between humans and the environment.

At this interface there are numerous factors, including population, socio-economic, political, biophysical, and cultural components that influence the lifestyles of community members and their interactions with the environment. Historically, India's policies, aid programs, and extension have dealt with these development issues through separate ministries, rather than approaching them from the viewpoint that people and the environment are interwoven.

Globally, the existence of this linkage between human health and the environment was acknowledged at the United Nations Conference on Environment and Development in 1992. The primary health care and environment millennium goals determined at this conference clearly state the international community's acceptance of this linkage between the environment and human health. Since then, efforts have been made to discover, design, and/or analyse initiatives and programs that combine community-based population and environment.

Though the global community and perhaps the Indian government have acknowledged the logic of combining health and environmental services, both are far from implementing such programs. Limited public resources coupled with burgeoning problems and dogmatic development strategies prevent more adoption of integrated community development programs in India.

The private sector has started to fill this neglected area, focusing on human development as an essential step towards sustainable and scaleable programs. The inherently interwoven framework of communities and human interactions with the environment attest that agriculture development is limited without human development.

Though agriculture development may improve the farmer's production, increase his income, and preserve the land for future generations, the sustainability of these results and the expansion to other individuals and communities may be compromised if health and socio-economic conditions are ignored and the people themselves are not taught self-reliance. There are several conceptualizations and methods of achieving human development that vary between organizations, but most share root convictions and the final goal is always self-reliant, confident individuals and communities capable of making informed choices and decisions.

According to Don Jose Elias Sanchez, whose work with poor farmers in Honduras is documented in the The Human Farm, human development grows from the conviction that only strong hearts and heads can guide the hands. Sanchez's training farm is centered around exercises and sessions designed to help farmers realise the strengths and understanding they bring to agriculture.

This ground-up, holistic approach develops human capacity in a working context that farmers can transfer back to their own communities. UNICEF on the other hand, perceives human development in three basic principles whose achievement varies across place.

The first principle is "sustainable human development evolves from a self-reliant understanding of local needs and resources." Communities have an intuitive sense of the dynamics of the local ecosystem, economic potentials, market forces, and social development variables. Recognizing these dynamics and giving community members the skills to collect data, respond to problems, and analyse situations improves self-esteem and can lead to a more sustainable project.

The second human development principle is "action must grow from a combination of a bottom-up and top-down programming." The informed decision-making needs to be based on a three-way inclusive dialogue between the entire community, government and agency officials, and experts. The third principle is "sustainability is only possible when action grows from community participation and self-reliance."

The success of an agriculture development project is contingent on the community's ability to analyse and respond to new problems after the withdrawal of the development agency. If communities are not taught the skills to understand and evaluate their own environment, the likelihood of expansion and/or sustainability is diminished. Here, in the third principle, is where the ideas of The Human Farm, UNICEF, and most human development schemes converge: the development of humans themselves empowers the individual and community, enabling the expansion and growth of their potential through their self-reliance and independence. This form of development is key to the future success of extension in India; the investment of time and training in people would assist the government in addressing the multitude of problems rural Indians encounter. The case study presented later in the paper elaborates on human development in the Indian context.

The integration of health and environment is advantageous to community-based extension because it addresses community interests as a whole, rather than as separate problems that are exclusively independent. Two of the foremost issues that need to be addressed in rural India are public health and agriculture. The leading causes of death in India include neo-natal, diarrhea, and pneumonia; infections and diseases that are all exacerbated by malnutrition.

Anemia, a micro-deficiency caused by lack of iron, is prevalent in more than 69% of pregnant women in India. Vegetarian diets, lack of protein and bio-available iron sources contribute to a deficiency that has severe consequences such as poor brain development in children, compromised immunity, maternal mortality, and reduced work capacity.

The majority of these deaths and illnesses could be prevented with proper health interventions and food supply improvements. The health of people is not contingent only on the availability of food, but the sanitary conditions and practices of families. Primary health care such as access to rural heath posts, education on proper infant and childcare, maternity care, reproductive care, portable water, and sanitation are all basic needs and rights the greater part of rural Indian people are denied.

Beyond the logical sentiment that this is iniquitous, researchers at the forefront of nutrition and economics argue child and adult malnutrition reduces labour output, GDP, cognitive ability, and motivation. It is not only beneficial to improve the health standards of people for their own wellbeing and ethics, but for the wellbeing of the country.

Agriculture and public health are linked in two ways. First, without sustainable farming techniques, diversified crops that bring in multiple nutrients, and income for supplemental foods, the consistent yield, sustained nutrition and capital of the farmer and his or her family are compromised. Second, poor health, .whether in the form of micronutrient deficiencies or parasites, can contribute to an inability to address farm issues with the energy and vigour expected from a healthy farmer and family.

This linkage of health and motivation is not only an example of the significance of linking health and agriculture, but it further validates the need for health in human development programs. Linking health and agriculture has the advantage of addressing these principal issues at the same time.

Furthermore, while it may be incongruent for experts or professionals to tackle health and agriculture together because they cross disciplines, it is natural for rural communities to perceive both as interconnected and integral to their personal growth. Communities are affected by multiple factors, and a holistic perspective is intuitive to them.

Advantages of holistic approaches to community development reach further beyond the more evident reasons discussed. This health and environment approach to community development can facilitate entry into communities. Projects that linked family planning and environment often found barriers were more easily overcome when environmental services provided the means of introducing family planning to the community.

Reproductive health is a potentially sensitive topic in India. These programs can generate confidence and community acceptance, in addition to being a way of creatively linking reproductive health to agriculture thus reducing potential hazards of discussing family planning. Environmental metaphors may include comparisons between the importance of spacing crops and the spacing of children or the need to prepare for children just as the land needs to be prepared for seeds. Additionally, the link has provided a gateway for women into the environment sector and men into the reproductive health arena.

Community-based population and environment programs can meet the needs of remote communities more efficiently than most sectoral, public sector, and advanced technology driven programs. The people that need health and environmental services the most are those living in remote areas with little access to roads, schools or health services. These areas are usually the last to receive government services, the least likely to accept such services when they are available, and the slowest to change attitudes and behaviors.

As a result, these areas in particular require strong community based holistic approaches to meet varied needs and improve adoption through the establishment of trust and the thorough understanding of community culture. Integrated and human development programs have the potential to actually generate human capital in these resource-limited regions.

This development of people is exceptionally important, as it is unlikely that further assistance will be readily accessible on the completion and withdrawal of the project. The strong base of combined health and environment programs gives the organizations the latitude necessary to establish this trust, understand the communities, and improve self-reliance in these remote areas.

There are other advantages to integrated development programs, including the inclusion of men and women in typically gender sensitive topics, cost effectiveness, and sustainability. The case study will demonstrate these advantages and those already discussed. The following section profiles the work of World Neighbors, an NGO presently working in India that has approached development holistically with an emphasis on health, environment, and specifically human development.

World Neighbors (WN) works in Karnataka, a dry land state in southern India. Though the state is considered at the median level of development in major sectors, the Human Development Report 1999 "describes the regions covered by WN programs as 'areas of darkness' with HDI indicators that fall below the sub Saharan African countries." The region is typical of many dry land areas around the country, with agriculture depending on scanty and erratic rains and over 90% lacking irrigation. As a result, farmers and shepherds of the region often migrate in search of food, fodder, and employment.

Malnutrition and infant mortality rates are high in comparison to other areas in the state. In 1998, WN began working in reproductive health in two villages; Nellur and Yelenavadgi, where agricultural activities focused on dry land agriculture, sheep health, and wool processing were already ongoing. Over the past three years WN has partnered with 8 additional NGOs to facilitate activities and trainings in order to meet several objectives:

- To organize target groups to undertake development activities.
- To create employment opportunities for the rural poor and establish training centers for self employment for youth and women.
- To plan and implement various rural employment and development programs with the cooperation of local groups for the all round development of the people.
- To sustain all development activities targeted to the groups.

As the objectives demonstrate, WN work in Karnataka focuses on human development in order to create sustainable, fruitful ventures, regardless if it's in the agriculture, health, or micro-credit sectors. One particular aspect of the programme targets the formation of women's groups through a three-step process. The first phase involves WN and the NGOs holding initial discussions with agriculture groups about the formation of women's groups in the villages.

Groups are established and reproductive health workers (RH) are identified and trained. In the second phase, the capacity building of women's groups with the help of the RH worker takes place.

The initial activities of the groups include the selection of leaders and training of leaders in such topics as group management, decision-making, recording keeping and monitoring and evaluation techniques. In the third phase, women's groups conduct activities independently, the RH worker visiting the groups only every three or six months.

The first two phases involve the intensive support of WN and the partner NGO to help establish the women's groups, develop capacity, and implement activities. These steps are essential to the successful development and coordination of the groups. Only when the skills and capacities of the groups are firmly established, about a three-year process, do WN and the NGO withdraw further and the women's groups conduct the activities independently.

Though the phases may be lengthy and intense, evaluation of the programme demonstrates that women are empowered in many aspects of their lives. An evaluation of the reproductive health component of WN's work was conducted in 2001, with indicators suggesting the human development methodologies improved the capacity of women to make informed decisions and the integration of health and agriculture was beneficial. Key findings from the evaluation state that "significant changes were reported in key reproductive health practices and rates for service use," compared to non-intervention villages. Additionally, "there were reported changes in indicators of women's status."

The women indicated that group activities improved decision-making, property ownership, girl child education, and reduction of violence against women." Women felt they were more confident "to organize and undertake a variety of initiatives ranging from closing bars and gambling houses to using civil disobedience to improve water supplies."

The linkage between men's agriculture groups and women's groups greatly contributed to the success of the women's groups. The agriculture group members assisted in the formation of the women's groups, encouraged their wives to attend, convinced other men to do the same, and supported them in terms of organization of events and discussions.

The linkage also encouraged communication between the men and women about dry land agriculture and "improved awareness of women's group members regarding key agricultural practices." Another advantage of the programme is the integrated model appeared cost-effective. Not only did the programme improve the group member's livelihoods, but spread to non-group members.

Finally, the inclusion of trainings in evaluation, monitoring, and implementation techniques helped ensure the sustainability of the programme. "The women's groups are confident in their ability to sustain key activities after the phase out of BSRDS." The development of human ability encourages the scaling up of a programme; it is a positive step towards the expansion of the programme to other communities.

The evaluation of the reproductive component of WN's work in Karnataka demonstrates the value of integrated programs. WN states on their Karnataka webpage, "the most important determining factor of success is the involvement and integration of various community groups." The apparent exclusivity of health and agriculture to some professionals is in direct contrast to rural community's instinctive perception that health, agriculture, and a myriad of other issues are innately associated.

Furthermore, the case study demonstrates that the development of human capacity enables communities to determine the focus of their projects and their own future. The development of human capital is one of the most efficient methods of any form of development as it empowers people to actually do the development themselves and enables communities and individuals to grow and expand beyond the original confines of their poverty. This form of development is part of a viable solution to the multiplicity of problems India faces.

"The processes through which past societies have undermined themselves by damaging their environments fall into eight categories: deforestation and habitat destruction, soil problems (erosion, salinization, and soil fertility losses), water management problems, overhunting, overfishing, effects of introduced species on native species, human population growth, and increased per capita impact of people."

This quotation from Jared Diamond's nonfiction book Collapse: How Societies Choose to Fail or Succeed reveals the profound impact human's management of the environment and natural resources has on the quality of their future existence. While the societies Diamond describes didn't necessarily disappear or spontaneously combust, their populations either declined or suffered monumental setbacks in their quality of life.

Furthermore, the societies did not collapse based solely on one of the eight categories, rather they collapsed based on the compounded affects of a combination of these categories. Four of the eight categories that undermine communities were discussed in this chapter: deforestation, soil problems, water management problems, and human population growth.

The occurrence of these issues in our paper and in Diamond's book is not a coincidence; the topics are acknowledged global concerns that must be addressed if we hope for an equitable, sustainable future. Our paper demonstrates that multi-functional approaches to environment and human development are integral to successful management in dry land regions of India.

Diamond's analysis of the collapse of societies is further evidence that if these issues are not tackled from an all-encompassing perspective, the triumph over one problem may be undermined by the eventual erosion of another. In the dry land region of India, where harsh environmental conditions interact with a growing population and little agricultural support, the reality of the situation merits the development of multi-functional extension programs.

WATERSHED AND DRAINAGE

Panchayati Raj is a system of democratic governance. It is an integral concept to the Indian political debate since Independence, with fervent arguments made both for and against the system. Until 1993 it operated sporadically and was included in the Constitution only as a directive principle. In 1993, forty-three years after Independence, the Constitution was amended and Panchayati Raj became part of the democratic system.

The Constitution specifies that there should be three tiers of local government: the village, the block and the district. The XIth Schedule, which accompanies the Amendment, specifies development areas over which Panchayati Raj Institutions (PRIs) will be responsible. Article 243 G (3) of the XIth Schedule is watershed development.

- These resources can also be developed under watershed development programs. Watershed development projects started to make a prominent appearance on the development agenda in the 1980s and 1990s. They were part of a variety of programs, such as Joint Forest Management, which started after the explicit policy recognition of the link between environmental degradation and poverty. The argument for a watershed approach was that it is an effective unit, in technical terms, for soil and water conservation. The acceptance that these physical boundaries may not coincide with human ones, whether defined geographically, socially or politically, has led to a spread in the scope of watershed projects. Watershed projects now commonly include a focus on wider rural development such as employment generation, credit, capacity building, afforestation and livestock improvement. The policy support for watershed development, the increase in funds for such development, and the widening scope of the programs themselves, led to a proliferation of projects.

- The Guidelines for watershed development were framed by the MRAE in 1994 to ensure that the approaches being used by the various programs complement each other.
- So far the Guidelines cover four government programs for which they outline a procedure for implementation which operates from the state, to the district, and down to the local body. The Guidelines specify what institutions should be formed at each level to monitor the programme, what their functions should be, and how the institutions should coordinate with each other.

The Guidelines are impressive in their scope, and in the emphasis that is put on local participation in the design and implementation of the programme. However there is one institutional area in the Guidelines which stands out as having very particular constraints and potentials. This is their procedure for the involvement of Panchayati Raj Institutions. The Guidelines specify that 'the Zilla Parishads (ZPs)/ DRDAs, as the case may be, shall be responsible for implementation of these guidelines at the district level'. Further, that 'the ZPs, Panchayat Samithis and the Gram Panchayats are also entitled to take on the responsibility of implementing a cluster of watershed projects in the capacity of project implementation agencies, if they so desire'. These provisions have a special significance because the Panchayati Raj Institutions are constitutionally part of a national democratic system. The ideology, mode of operation, and objectives of PRIs differ from those of the district administration.

The Guidelines, by using clauses such as 'as the case may be' and 'if they so desire', remain vague about responsibility, roles and coordination. The implementation of Panchayati Raj varies from state to state, within the broad parameters specified by the constitution. In some states, PRIs have been given authority at the district levels, and considerable power to wield this.

However in most of the states the relations between Panchayati Raj Institutions and the district administration hangs in a balance. Despite their constitutional importance, the former often lacks political and administrative support as well as funds. This is the context within which the Guidelines will be implemented. The uncertainties of the Guidelines are not, by definition, a problem.

There is a possibility that the uncertainties can be resolved at the district, block and village level to the mutual benefit of the two systems. However, there is also abundant empirical evidence of conflict between the district administration and PRIs, which impairs the development of the systems, especially PRIs, and dilutes the effectiveness of development programs.

THE TERMS OF REFERENCE AND RESEARCH METHODOLOGY

This report investigates the links that exist between the administrative and political institutions in the Guidelines. Although this report focuses narrowly on the institutions as they relate to watersheds, the issue of political administrative links is much broader. I will draw on some generic issues and wider empirical evidence where it is useful, for an understanding of watershed development.

The ultimate objective, of which this report forms only a small part, is to isolate means by which the two systems can complement each other to ensure effective local development. There are four particular areas that this report will investigate:

- The philosophy and objectives of these bodies with respect to watershed development.
- The nature of the linkages – both formal and informal – between Watershed Institutions (WIs) and political bodies.
- The difference in roles and modes of operation between the two bodies and how the balance of power between these is established.
- The mechanisms that exist for ensuring the transparency and accountability of these institutions at the district, block and village levels.

This report cannot hope to be exhaustive in the exploration of these questions. There is too much variation and too little time for such an ambition.

It is also too early; the Guidelines have only been effective since April 1995 and the Constitutional Amendments since 1993. Instead, the intention is to isolate broad differences in opinion and preliminary evidence on the questions above which will assist in future policy and implementation. Both systems are in a transitional phase; this report is a contribution to this process.

RESEARCH METHODOLOGY

The report starts with a brief review of the empirical experience of cooperation and conflict between Panchayati Raj and the state bureaucracy. Panchayati Raj has, since Independence, excited great differences in opinion. I examine these and consider some of the conceptual issues around which they revolve. For this I draw briefly on some of the literature on decentralisation, institutional change and collective action.

The research was conducted in Karnataka and Andhra Pradesh. The federal structure of India, in which states have considerable autonomy to adapt legislation and institutions, makes generalisation very difficult. However, because it means variations within a common framework, comparisons between states can be very useful. Karnataka was chosen because it has been progressive in the implementation of Panchayati Raj prior to the 73 rd Constitutional Amendment.

It has also been comparatively progressive in engaging with the provisions of the 73rd Amendment. Andhra Pradesh was chosen because it has a comparatively effective bureaucracy, but has not given as much support as Karnataka to the implementation of Panchayati Raj. The bulk of the research is based on extended interviews with those involved in the implementation of the Guidelines.

This includes government officials in MRAE in New Delhi; state government officials; donors; members of Zilla Parishads and Gram Sabhas; Non-Governmental Organisations (NGOs); Project Implemention Agencies (PIAs); Watershed Associations (WAs); and academics doing applied work on Panchayati Raj. 4 The questions explored in detail, what the interviewees considered to be strengths and weaknesses of the political-administrative link, and how and when these could be practically resolved.

The exploration of this detail was important because there is a general consensus that both Panchayati Raj and state planning are necessary. In this case the clauses in the Guidelines as the case may be and if they so desire do not constitute a problem. The differences in opinion emerge in considering the practical resolution of questions of authority and power.

The purpose of this section is to illustrate that the ideal juncture between Panchayati Raj and the state administration is a subject of many debates. There is much which can be learnt from past experience on this subject. The section ends with an isolation of the key conceptual and practical research questions. It will also detail some general features of development planning in India and the context within which the Guidelines were introduced. The objective in relating this history is to show that the development of Panchayati Raj and the instruments of state planning are closely linked. Political decentralisation has often been accompanied by an increase in the functions and power of the state bureaucracy. A consideration of the past swings between centralisation and decentralisation, and between Panchayati Raj and the state, planning is therefore useful in understanding the context of the Guidelines.

PANCHAYATI RAJ AND THE 73RD CONSTITUTIONAL AMENDMENT

The history of Panchayati Raj can be divided into three phases. The first phase began in 1959 when the failures of state planning made it necessary to establish local bodies to ensure people's participation in development. Panchayats, for reasons I will outline, stagnated from around 1964 until the second phase began in 1977 with the appointment of the Ashok Mehta Committee. The Committee was given a mandate to find ways to revitalise Panchayati Raj.

The recommendations made were intended to give Panchayats a new lease of life as democratic bodies. The third phase is the one that we are now in which is marked by the passing of the 73rd Constitutional Amendment. The first and second generations of Panchayats were established with different principles and objectives. The first generation was established as a local body to ensure people's participation in development programs.

The priority was for Panchayats to be institutions for development. In the second generation the emphasis was on establishing Panchayats as democratic institutions and local units of self-government. The third generation has enshrined both principles in the constitution and therefore has the potential to be the most 'revolutionary' and perhaps end the phases and establish Panchayati Raj Institutions as a stable feature of Indian democracy.

An assessment of this potential can usefully start with an exploration of the reasons for past failures and successes. Panchayati Raj Institutions, despite Gandhi's insistence that they were central to a restructuring of India, were only included in the directive principles of the Constitution which is not justifiable. The process of development was entrusted to the Community Development Programme (CDP), administered by the state, which was introduced in 1952. The first phase of Panchayati Raj began in 1959 due to the failure of the CDP to enlist local participation.

The Balwantrai Mehta Committee, set up to explore how to redress this situation, recommended that local participation should be encouraged through statutory representative bodies which provided local people with a feeling of ownership over the projects. By 1959 all the states had passed Panchayat Acts and the system had spread throughout the country by the mid-1960s. By the early 1970s it was on a decline. There are varied opinions on why this was so, which I will consider in more detail in various parts of the report. There is a broad consensus on two issues.

The first area of consensus is that the state administration and state level politicians felt threatened (and still do) by Panchayati Raj Institutions. PRIs, if they fulfil their mandate, have the potential to make many of the functional roles of the state administration redundant.

There is therefore, a direct conflict between the state administration and PRIs. State level politicians feel threatened by PRIs because they have the potential to erode their power base. Observers point to a deliberate plan by the bureaucracy to retard the growth of PRI, by launching state development programs without going through the Zilla Parishads. Funds to the Zilla Parishads were slowed down and Panchayats as agents of change were replaced by state administered development programs.

The second area of relative consensus is that most Panchayats had by the 1970s become corrupt power bases for local élites (and still are). Elections were not held, allowing those who had come into power to entrench their positions and create links with state level political leaders. The centre was more interested in the vote banks created through this system than in taking any action to ensure that PRIs were democratic. The differences in opinion lie in the achievements of Panchayati Raj during that period, and what would have happened if they had had state support.

Mathew (1995), whilst admitting that many Panchayats were corrupt, argues that there are many official reports during the mid-1960s which show that there were also positive developments. These reports reveal local recognition of the value of Panchayats, the emergence of competition for local leadership, and the successful implementation of schemes, especially for education.

Most critically, Mathew argues that the domination of Panchayat by élites 'is essentially a socio-political problem and could have been tackled to a great extent by holding elections at regular intervals. The resulting political education of communities, oppressed for ages, would have certainly changed the scenario'. This same question, of whether democratic practice can bring development and change local power bases, is at the heart of the discussion over PRIs today.

It is also the main issue that emerges, (along with practical project boundaries) in comparing the potential of WAs and Gram Sabhas. The second phase of Panchayati Raj began in 1977 with the appointment of the Ashok Mehta Committee.

There are several reasons for a renewed focus on PRIs. One was the emergence of regional parties representing regional concerns on the national agenda. States began to argue that the centre had dominated the development process and had not respected the autonomy of the states as is required by the Federal Constitution. For example, the Karnataka Minister for Panchayati Raj, Abdul Nazir Sab, demanded a 'four pillar state' based on genuine federalism and the establishment of local Panchayats. The Ashok Mehta Committee recommended that PRIs be genuine units of self- government, not merely institutions that execute state centred development programs. The Government of India (GOI) recognised the political necessity of decentralising, and appointed various committees during the 1980s to explore Panchayati Raj.

The experience of the second phase of Panchayati Raj varies from state to state. It is generally acknowledged that the most earnest efforts were made by West Bengal, Karnataka, Andhra Pradesh and Kerala. The Karnataka experience in particular has been recognised as revolutionary with a new form of district administration in which the power is devolved to Zilla Parishads, with the Adhyaksha (President) having the status of the Minister of State.

However the GOI has continued to doubt the ability of PRIs and has simultaneously rearranged administrative structures at the district level and below to administer centrally sponsored development programs. Every new scheme has come with new agencies, and the business of ensuring public participation has been entrusted to voluntary organisations and NGOs.

The issues that have influenced the evolution of PRIs in the first two phases are still essentially the same issues that are being debated after the 73 rd Amendment. These issues revolve around whether PRIs can manage development, and the appropriate division of tasks between the state and PRIs. Almost all government representatives, with notable exceptions, were sceptical of the ability of PRIs to implement the Guidelines. There are two main differences in the current discussions. One is the presence of NGOs in the equation, and the differences in opinion over what their role should be. The other is that the centre is less stable and increasingly has to concede to regional demands, including the demand for a decentralisation of power.

STATE PLANNING AND THE GUIDELINES

India adopted a strategy of state-led capitalism after Independence in which the state justified its role as directing a programme of development on behalf of the nation.

The educated professionals, as leaders of a consensus nationalist movement, took the responsibility of planning development and assumed control of the state apparatus. They depicted planning as a rational way to pursue policies outside the political process. State planning provided for the flow of authority from the central government to the state and from there to the district, block and village. The local development bureaucracy expanded enormously after the CDP was started in 1952.

The main justification for not fully adopting Panchayati Raj, was that development has to proceed without being politically corrupted. The state did recognise the practical and political need to involve local people in development and made sporadic moves at establishing Panchayati Raj. However because elections were rarely held, and PRIs were not given funds, they became bodies which were dependent on the government for their survival. Chatterjee has noted that 'a developmental state operating within the framework of representative politics would necessarily require the state to assume the role of the central allocator if it is to legitimise its authority in the political domain'.

From the sixth five year plan onwards, the planning process became increasingly centralised and programs and the selective distribution of these benefits became part of the process by which the GOI retained political control. Many of the poverty alleviation programs started during this period and were directed straight to the District Rural Development Agencies (DRDAs).

These are registered societies under the Collector, and were established for the central purpose of implementing centrally sponsored programs. The Gram Panchayats became part of the delivery system, and a means through which to harness votes. Most programs established their own institutions at the village level for the implementation of the project.

The 1980s and 1990s have seen an enormous growth of local level institutions established by the government. Community based Organisations (CBOs) and Self-help Groups (SHGs) often carry the name of the project: Women and Child Committee, Village Forest Committee, Watershed Association etc. The central government has been under increasing internal and external pressure to decentralise since the 1990s. The external pressure had come from rising balance of payments problems that led the central government to seek external financial assistance and accept the combined international development agenda of liberalisation and governance. The involvement of NGOs in implementing development programs in partnership with the state has been a central part of the agenda.

The Eighth five year plan recognised the 'need to decentralise control over nature and natural resources', and to involve voluntary organisations and NGOs in this process. The internal pressure to decentralise comes from state governments and regional movements. The 73 rd Constitutional Amendment is a manifestation of the political need to incorporate these demands.

The congress (India) proposed these Amendments because of the patent popularity of non-congress parties and because the dangers of over-centralisation were becoming increasingly obvious. This is the context within which the Guidelines were framed and are being implemented. The Guidelines specify a line of coordination for watershed projects from the central government to the village. All of the conflicting pressures for centralisation and decentralisation are manifest in the Guidelines. It is basically a centrally sponsored (with state contributions) Poverty Alleviation Programme that operates directly at the district.

There is an unprecedented move to give local Watershed Committees (WCs) control over the programme. However centralised control over these is still a problem. The Guidelines specify that PRIs should be involved but the legislation to do so has not yet been put into place. NGOs have been centrally involved in implementing the Guidelines but their role, and how they relate to the other stakeholders, is unclear. The resolution of these issues is heavily influenced by particular constellations of power in different states, and has to be considered within the broader perspective of the policy environment for development programs.

THE RESEARCH QUESTION

The immediate question of the links that exist between institutions established for watershed development and the Panchayati Raj system is part of the broader question of how the political and administrative system should work together.

The implementation of this institutional model is complicated in practice. The Guidelines operate on ideal types and do not address existing areas of conflict and cooperation between institutions. All interviewees thought that the Panchayati Raj Amendments are a step in the right direction, and that PRIs should ultimately be responsible for implementing development programs. The differences in opinion, and debate, emerged in discussing the principles and practice of decentralisation, institutional change and collective action.

First, what kind of decentralisation is necessary for watershed management? Although all of?the interviewees thought that PRIs are important, many also thought that their involvement in watershed management in the near future was not practical. They argued that we should concentrate on effective administrative decentralisation to the institutions created by the Guidelines. The difference in opinion is essentially one over whether or not democratic decentralisation can be a means by which to achieve development.

Are the people ready for Panchayati Raj or will it be captured by local élites? The second question is whether the implementation of the Guidelines, and the involvement of PRIs, requires institutional change. The Guidelines refer to PRIs, Watershed Institutions, the government and NGOs as ideal types.

These ideal types are however based on already existing institutional alliances and conflicts. For example; alliances and conflicts between sectors (NGO, government); within the government (centre, state and district); and between PRIs and the government and NGOs. Is it necessary to change the underlying relations between these institutions for the Watershed Guidelines to be implemented?

If yes, what role will the different institutions play and will there be winners and losers? The third question is how to stimulate and support local collective action. The existing state administration and PRI stimulate links of patronage and clientelism that lead out of the village or Gram Sabha. How can social capital be built which enables local decision-making and demands state and PRI accountability?

How do the Guidelines and PRIs, and their treatment of the community, gender and caste, compare in their potential to achieve this? Critically, do political structures encourage or retard the growth of local participation and awareness Political ideological differences on these issues are manifested in the implementation of the Guidelines in practice.

Panchayati Raj and the state administration differ in their vision of decentralisation; in the institutional change they propose, and in how to organise collective action. In particular the routine interaction between the DRDA and the ZP; the practical resolution of boundary problems (physical, social, political); how PRIs and Watershed Institutions compare in their treatment of disadvantaged sections of the community; and what systems exist for monitoring and accountability.

The PRIs and state administration are in a period of transition. In the case of the latter this is inevitably true, but it is particularly evident now with the breakdown of central authority, and the emergence of regional state agendas for development. This is manifest in the state wise implementation of Panchayati Raj and the Guidelines. The respondents all wavered between cynicism about the potential of Panchayati Raj, and optimism that the transition was in the right direction.

The Guidelines, despite the rigid structure that they lay down, leave much of the identification of authority open between hierarchies. There is a danger that this will simply mean a continuation of the status quo. However there is also the potential to make the transition to a more decentralised and inclusive development programme, through both the political and administrative system. As Gandhi said of the realisation of Panchayati Raj 'We must have a proper picture of what we want before we can have something approaching it'.

DECENTRALISATION

Decentralisation of development planning and implementation are central to both Panchayati Raj and the Guidelines. It is important therefore to examine the compatibility of the definitions of decentralisation, development and governance embodied in both systems. There are two main objectives of decentralisation: one is to enable communities to meet their basic needs to ensure physical well being and the other is the realisation of civil and political freedom, democracy and autonomy. Whether these latter are included as an explicit purpose, and if so, whether they are included as fundamental rights or simply as laudable goals, affects the political character of the decentralisation programme.

These questions address some fundamental issues: what counts as development? What role should the state play in development and what criteria should confine their areas of control? The practical resolution of these issues in India has been especially influenced by two factors. The first is pressure from state governments and sections of the electorate to devolve power.

The second is pressure from international donors for the state to scale down direct implementation of development programs. Donors have insisted that India decentralises management and works towards good governance, described as 'the manner in which power is exercised in the management of a country's economic and social resources'.

The agenda for good governance includes a reformed public sector to ensure efficiency and transparency. Development programs should be implemented through cooperation between the state and institutions in civil society, especially NGOs. The logic behind the construction of these partnerships is that both the state and NGOs have comparative advantages which combined, can improve the delivery of development programs.

In particular, NGOs can ensure that projects stimulate 'participatory development' because they operate at a smaller scale, slower pace, and are more flexible and attuned to local needs. State withdrawal in India is still slow, but the importance of involving NGOs was centrally accepted in the seventh five year plan and has been manifested in the rapid proliferation of NGOs during the late 1980s and 1990s. It is especially evident in programs on poverty and the environment such as those for watershed management.

The governance agenda has a powerful influence on the way in which democracy and development are conceived in India. The governance agenda recognises the principle of democratic rights and the importance of accountability. However there is little consideration in the governance agenda of the role of democratic politics in deciding the principles of governance. 'The question of whether demands for governance – good, or efficient, or equitable – should emanate from the citizens of a polity, and exclusively from them, is not generally raised'.

The practical relevance of these questions was manifested in the discussions that I had with respondents over the implementation of the Guidelines. What relation does governance (as exercised through the Guidelines) have to democracy and politics (as manifested in Panchayati Raj)? In practice, the state administration takes the decisions on development, and these are implemented by the Guidelines. The exclusion of politics results in a restrictive definition of democracy as defined by the formal institutions and procedures. PRIs are included in the system not as a check and balance on accountability, or as part of the procedure for making policy decisions, but as an implementing agency. There are several implications that can be drawn from this interpretation of democracy and development.

The first is that development (and the system of governance to implement it) is not open to political contestation. The second is that good governance can ensure accountability and transparency. But to who? If institutions of governance (such as the Guidelines) are not accountable to the democratic polity (PRIs) then to who are they accountable? Third, and conversely, if the formal conditions of democracy are fulfilled (73 rd Constitutional Amendment), does 'good governance' necessarily follow? As Lateef observes, 'accountability exists in the formal sense in India.

We have the option to "throw the rascals out" and frequently do. But accountability is eroded by the lack of true choice for voters'.

A democracy that ensures good governance, and accountability to voters, has to be politically contested and constituted. This conclusion was abundantly clear in the operation of the Panchayati Raj system in Karnataka The central question in most development projects today, which includes those for watershed management, is not how to work with the democratic system, but how to generate 'participation'.

Participatory development describes the principle that people have to be part of the identification of their development needs and problems, and have to be part of the design for the solution. The concept of participation has been much studied and different 'depths' of participation have been identified, ranging from 'passive participation' to 'self-mobilisation'. Participation is now identified, arguably, as the most important element in development programs, and the lack of participation the reason that development programs fail.

There is no doubt that participation is important: but how do participation in democratic development and participation in development projects compare? An assumption of the participatory logic of the latter is implicit in the participatory logic of the latter is the assumption that the root cause of social problems can be located in local social attributes, rather than the structures, property regimes and social orientation of the state.

Participation is intended to empower local people and enhance project effectiveness by ensuring that people have a stake in the project. Participation in this context is a component of a project and confined to implementation. The participatory process does not influence the definition of priorities at the district, state or national level, and as a component of an externally assisted programme, it is insulated from formal political processes.

The notion of what decentralisation entails, and how communities should participate, is a major difference between the Constitutional Amendment and the Guidelines. The usual reason given for not working through PRIs is that they are governed by élites and too corrupt. This logic assumes that governance can bring development, and that development should precede democracy. As Jayal notes a person becomes a 'citizen-in-the-making', 'a subject: a passive recipient of rights, enjoyer of governance, beneficiary of development'. Whilst this option sounds unattractive, there is plenty of evidence from East Asia that the benefits of state administered development can be substantial.

There is also plenty of evidence from the early days of Panchayati Raj that the human costs of challenging power structures at the local level can be hard. I have drawn the lines between participatory development (as envisaged in the Guidelines), and democratic development (as provided for in the constitution), sharply for illustrative purposes.

The difference is tangible. However in practice, both government officials and NGO interviewees saw the boundaries between the two as more fluid. They all agreed that democratic development is the final goal and that building local capacity could lead to this objective in a less confrontational, more efficient manner. Local opinion on how to do this varied, as I will go on to discuss.

DECENTRALISATION IN PRACTICE

The discussion above considers the 73rd Amendment and the Guidelines in the 'spirit' in which they were written. The difference between 'the spirit of the law' and practice was a recurring theme in all the discussions. The Guidelines and the 73 rd Amendment are both progressive on paper. However it is important to recognise the loopholes which provide escape routes from transferring power and authority to local bodies.

The Guidelines provide for devolution of decision-making power from the central government down to the local level. They are also considered to have the potential for stimulating 'a tremendous change in the attitude of government functionaries...[and so]...a quiet revolution in rural India'. The 73 rd Amendment provides for units of self-government that will fundamentally and permanently alter the democratic process. This section takes a closer look at the legislation on each system and then considers the implications of these for decentralisation in general and the implementation of the Guidelines in particular.

DECENTRALISATION IN THE 73RD CONSTITUTIONAL AMENDMENT

The 73rd Amendment has been widely welcomed as constituting landmark legislation which provides a constitutional basis and continuity to the Panchayati Raj system. It specifies that Panchayats shall be established throughout the country at all three tiers, that elections shall be held every 5 years, that one third of all elective seats shall be reserved for women and about 15-- 18% for scheduled castes and tribes. It has specified a developmental role for Panchayats at all three tiers and a District Planning Committee to coordinate these.

The 29 areas over which the PRIs have jurisdiction are provided for in the eleventh schedule of the constitution. A state Election Commission has been established to supervise Panchayat elections and a state Finance Commission to make recommendations about the principles that should govern the distribution, sharing and raising of revenue.

The state governments have been given considerable freedom over how the 'spirit' of the Amendments is to be implemented. These are specified in the Conformity Acts that the states are required to submit to the central government.

As a long time observer of Panchayati Raj commented 'PRIs have now become an inalienable part of the Indian polity. We cannot work around them any longer. They are here to stay.' These are all positive developments. Unfortunately there is a considerable gap between the spirit of the legislation and the manner in which it is being implemented. In almost all states the implementation of Panchayati Raj has gone through phases of focussed attention and decline.

Political parties often disable PRIs established by a previous government in order to disable what they perceive to be local power bases of that government. Political parties also frequently promise Panchayati Raj as part of their election manifestos, and then establish these in some form, however diluted, when they are elected. Both these swings are apparent in Karnataka and Andhra Pradesh. Karnataka passed a very progressive Panchayati Raj Act in 1987 under the Janata Dal government that attracted attention all over the country because of its radical provisions for the devolution of power to districts and villages.

The general approach was that any activity whose area of benefit did not extend beyond the district should be managed at that level. The DRDA was abolished and merged with the Zilla Parishad, which was given wide-ranging functions. Detailed orders were given by the state government specifying the programs, personnel and funds of each department that would be put under the Zilla Parishad. The Adhyaskhsa was given the status of State Minister. In Andhra Pradesh the Telugu Desam Party made a reconstruction of Panchayati Raj system part of their election manifesto in 1984.

They created a third tier that brought 'government closer to the people' and held elections in 1986. There has been some progress in both states to implement the recommendations of the 73rd Amendments. However this is a slow process and many observers have commented that in both states PRIs are actually less decentralised, autonomous and progressive than before.

In Karnataka the congress government rolled back many of the provisions of the earlier act in the early 1990s. The remaining part of this section will focus on some problems in the implementation of both the spirit and the letter of the law, which are common to most states. The broad areas of concern are:

- The lack of functional autonomy and poor identification of what constitutes self-government?
- The lack of financial autonomy?
- Lack of legal authority. Underlying all of these is political and bureaucratic opposition to the?full-scale decentralization of power

Functional and Administrative Decentralisation

Although the Constitution identifies Panchayats as institutions of self-government, the scope of 'self-government' has not yet been defined. The state governments have been given the freedom to decide on the operation of PRIs and many have chosen to interpret self-government as 'agency for the implementation of development programs'. The accepted norms of decentralised self- government are:

- A function that can be done at a particular level should be done at that level and no higher.
- The allocation of functions at the various levels should complement each other.
- the line of command should be clear so that the autonomy of each tier is observed.
- PRIs can never be units of self-government if they continue to function as agents of the government.
- Local level planning of local resources should be an important function of Panchayats, and?should be backed by financial and judicial powers (Institute of Social Studies, 1995:19)

In most states decentralisation has been limited to development works on the 29 subjects listed in the XIth Schedule, and even in this narrow domain the transfer of authority has been haphazard and the division of powers unclear. The state government retains the power to assign or withhold functions, which is constitutionally a job for the state legislatures. The bureaucracy has been reluctant to frame the rules for decentralisation and most poverty alleviation projects still flow through the DRDA.

Watershed development, which is on the XIth Schedule, is a case in point. However in Karnataka, the process of decentralisation has gone further than in most states. Watershed projects are run by the Chief Executive Officer (an administrative officer) who operates from the Zilla Parishad of which the Adhyaksha, an elected representative, is the President.

In Andhra Pradesh watershed projects are presently run through the DRDAs. Financial autonomy Functional autonomy has little meaning unless it is backed by financial autonomy. PRIs have survived mainly on 'grants-in-aid' from the central and state governments. Central funds for specific schemes are meant to flow directly to PRIs (through DRDAs if they exist) with Panchayats acting as the delivery agents for these schemes. In addition Gram Panchayats receive an annual grant from the state government.

Under the 73 rd Constitutional Amendment Gram Panchayats are eligible to raise revenue and to a share of state revenue. However the details of this are left to the state Finance Commissions and most of these have yet to specify their conclusions. Commentators have argued that the devolution of revenue-raising powers to the Gram Panchayats is critical.

Rajaraman et al. argue that fiscal devolution should be linked to the functions which are listed in the XIth Schedule. These subjects, such as levies on water resources, and royalties on the use of common property, constitute the 'natural domain' of local government. However presently they are being collected by the state. The approach paper to the ninth five year plan states that 'the Panchayati Raj Institutions must mobilise local resources and the State Act must empower them to levy taxes and cesses at different levels'. Further that 'in the case of Centrally Sponsored Schemes, it is proposed that during the ninth plan the flow of funds would be directly tied to the Panchayati Raj Institutions'.

The details of how to assign tax duties and revenue-raising powers to different tiers of the PRIs is a very complex issue. Many commentators argue that it is a vital starting point for autonomy. Revenue collection from; water, forests and other common resources are not so complicated and have the potential to provide large revenue to the Gram Panchayats.

The ability to raise revenue, or to collect public contributions, is important not only as a source of income but as a means through which to promote participation. It is also a means through which to promote accountability to local people. *Judicial autonomy* Finally, the transformation of PRIs into units of self-government is hindered by their lack of judicial autonomy. Local self-government without any policing or judicial powers lack credibility, and therefore one of the provisions that many observers have stressed is essential are Nyaya Panchayats; a system of decentralised justice.

DECENTRALISATION AND THE GUIDELINES

The Guidelines provide for a devolution of decision-making powers to the district and for the transfer of funds from the central government, along with the share from the state government, to the district, and from there directly to the Watershed Committees. As such it is no different from most of the poverty alleviation programs which run through the DRDA.

The Guidelines also specify that the Watershed Committee should open a bank account for the receipt of money for watershed development. Another interest bearing account is to be opened for money generated by the project in contributions, user fees etc., to be used for post-project watershed maintenance. Both funds are under the management of the Watershed Secretary, a member of the Wasteland Development Team, and the chairman of the Watershed Committee.

This provision for decentralisation of funds is new. So is the emphasis that the Guidelines place on cooperation between government institutions, NGOs, and PRIs. The extent to which the Guidelines are able to ensure decentralised watershed development is limited by several factors. First, the Watershed Secretaries, PIAs, watershed development Trusts (WDTs), WAs, and WCs are all approved from the hierarchy above them.

All of these institutions are dependent on the hierarchy for funds and are in the final instance accountable to the government. The tendency for the Collector to retain control over decision-making persists in the implementation of the Guidelines. 'Collectors continue to retain control over decision-making processes, rather than moving towards a more facilitative and supportive role emphasising a decentralisation of power and responsibility'.

The extent to which the Guidelines can ensure watershed management can only be judged by what remains in terms of physical or institutional capacity and structures once the project withdraws. The Guidelines specify throughout that the PR institutions can, may, and should be involved in the implementation of watershed development programs. However without specifying their areas of authority the Guidelines leave the operation of the Guidelines to the status quo in power relations, and in most states the upper hand is held by the state administration. There are several areas in which immediate questions arise. The Guidelines (article 39) provide for the establishment of a District watershed development Advisory Committee (DWDAC).

The 73rd Constitutional Amendment (Article 243 G) provides for the establishment of a District Development Committee (DDC). The XIth Schedule lists watershed development as one of its areas of control. Is this a duplication of activities? The Guidelines state that 'the DRDA/ ZP, as the case may be, shall be responsible for the implementation of these Guidelines at the district level.' (article 29). The ZP can be responsible for the implementation of the watershed programme 'once they come into operation with adequate powers and resources'.

Where the DRDA has been made responsible, the Chief Executive Officer will be a member of the advisory committee. 'The Zilla Parishad at the district level and the Panchayat Samiti at the block level shall have the right to monitor and review the implementation of the program' (Article 29). As I will go on to discuss in the report, given the past experience of working relations between PRI and the administration (an experience which is being currently repeated) these provisions are very vague and will not change the status quo.

Is are not legal or statutory bodies and there is considerable uncertainty over their legal rights over renewable natural resources and the contractual arrangements that should govern the relations between them and the owners (state governments or PRIs). WIs have the right to register themselves, and are encouraged to do so in the Guidelines. Sporadic efforts have also been made to federate these groups at the district and sub-district level.

As independent organisations these groups can federate at whatever level they want to, but if they are supported by development organisations and donors then they may conflict with higher level PRIs. The PRIs have been created for decentralised development so perhaps there should be more of an explicit appreciation of their ability to provide the umbrella institution. The Karnataka Panchayati Raj Act through an Amendment in 1997 provides for committees to be set up by the GP for specific purposes with provisions to have non-elected persons to be members.

This arrangement has been provided to fill the gap that existed at the village level of any institutional arrangements to involve the community in development work. The ninth plan provides for the provision of funds for centrally sponsored schemes to go?directly to Panchayati Raj Institutions. Interviewees from the Department of Rural Development (DRD) in New Delhi confirmed that this was being considered. How will this be coordinated with the funds going for watershed development? What is the position of the state governments on this? All of the interviewees from state governments saw the devolution of funds to the district level (whether PRI or DRDA) as, in the words of one interviewee 'a vote of no- confidence against the state'. They considered it to have detrimental affects for the evolution of decentralised government. I will return to consider this issue again in some detail in later sections

There are several conclusions that can be reached from considering the history of decentralisation in India. The first is that a common way to avoid implementing the spirit of the law is to pay a lot of attention to the letter of the law. A respondent in Bangalore commented about Panchayati Raj 'the law itself is a shifting target. How can we possibly implement it?' Without specific measures to implement laws on decentralisation, it is likely that the status quo power will continue to dominate, and in most states this is held with the bureaucratic and the political élite.

The publication of the 73rd Amendment is therefore a necessary but not a sufficient condition for decentralising democracy. The second related conclusion is that it is important to deconstruct what is meant by decentralisation and not to confuse the roles of the administration and the political system. An efficient bureaucracy that can deliver services and goods continues to be important in a decentralised political system. However decentralisation without a transfer of authority and autonomy is effectively 'passing the buck along' and diversifying opposition to central control.

Chapter 9

Institutional Change, Cooperation and Conflict

The respondents in this study often commented that India today is in a period of institutional transition. There is a widespread recognition that the state, and in particular the central government, has 'failed' in planning development, and that a broader institutional alliance is now critical for this purpose. The Guidelines are a representative reflection of this transition and change in approach. They provide for collaboration between the government, democratic institutions and NGOs. This section will look at the linkages, both formal and informal, between these institutions.

The guiding questions are: what are the determinants of institutional change? What comparative advantages do the different institutions have for watershed development? What are the areas of conflict and collaboration? If we follow the 'spirit' of the 73rd Amendment and the Guidelines, which institutions, if any, have to change and how?

The pressure for institutional change in development planning and implementation, as noted in the last section, has come both from international donors and from internal political and economic pressure. Within Indian State Governments, regional movements, and an increasingly mobilised electorate have put pressure on the central government to decentralise. International donors have at the same time placed pressure on the government to continue economic liberalisation and to scale down the public sector. The emphasis is no longer on state withdrawal (as it was in the early 1990s) but on public sector reform and capacity building. This change in perspective is partly due to revisionist perspectives on the role of the state inspired by development in East Asia and the role of state bureaucracies in enabling this.

Empirical and conceptual work arising from the new institutional economics framework has also challenged the dominant role ascribed to the market during the 1990s. The general trend has been to a reconsideration of the role of the state, rather than prescribing its withdrawal. 'Matching role to capability involves not only what the state does but also how it does it. Rethinking the state also means exploring alternative instruments, existing or new, that can enhance state effectiveness'.

These ideas provide the official context of development policy and the operating principle for donors. In India the importance given to institutional cooperation in development planning is manifested in the eighth and ninth five year plans and in the rapid growth of NGOs involved in poverty alleviation projects throughout the 1990s. The rethinking of the role of the state has led to a large literature on cooperation between the state and civil society institutions. The role that political institutions should play in this collaboration is less evident. The importance of 'democracy' and of social movements is generally recognised but the practical role of politics or of local political institutions remains unexplored.

This policy failure to examine the role that political institutions should play in the development programs is reflected in the discussion around the implementation of the Guidelines. The theoretical and practical concentration has so far been on the role that the government should play, where and how it should cooperate with the voluntary sector, and in which areas there is a need for capacity building.

However, as White argues, an effective crafting of institutions requires us 'to think systematically about the political dynamics of democracies in developing societies'. This section will first look at the arguments made for state-society collaboration in the literature and in practice. The importance given to political institutions, and their potential role, can be deduced by what the literature assumes, and what it leaves out.

COLLABORATION BETWEEN THE STATE, NGOS AND CIVIL SOCIETY INSTITUTIONS

Most recent poverty alleviation projects in India have emphasised the importance of involving both NGOs and local civil society institutions, such as CBOs or SHGs. The Guidelines for watershed management are typical in this respect. The rationale for NGO involvement is that: they operate on a smaller scale; have a better understanding of local priorities, and are more flexible in their methods of implementation. They are able to synthesise practice and policy because they are embedded in both worlds.

These comparative advantages hold true for watershed management, as an exploration of the performance of NGOs and the government as Project Implementation Agencies has established. These arguments about the comparative advantages of NGOs in the implementation of projects have been expanded to a broader consideration of state-society synergy.

The principle is 'that active government and mobilised communities can enhance each others developmental efforts' on state-society relations has identified two underlying processes that contribute towards synergy. The first type are complementary relations between the state-society where putting together the resources of both results in a greater joint output than either can deliver on their own. In watershed development the classical division would be the government providing the technology, the people providing the labour and the local knowledge and the NGOs facilitating the interaction between the two.

The other type of relation is one of embeddedness; the notion that state-society relations are affected by ties that cut across the public-private divide. Such relations have conventionally been considered in terms of the hazard of corruption and rent seeking. Empirical studies have shown the beneficial affects of such ties by increasing accountability and building on behavioural characteristics such as trust and obligation. The seminal studies on the operation of complementarity and embeddedness in the context of development projects and programs have been done on irrigation departments and their clients. Like those departments working in watershed management termed as 'street level bureaucracies'. The operators of these departments:

- Interact daily with citizens-clients in the course of their jobs.
- Have substantial discretion in the execution of their work.
- Use their discretion in ways that are not easy to monitor, in the sense that it is difficult to identify precisely what they do.

Wade (1993) has compared the operation of the irrigation department in India and South Korea and tried to understand why the latter operates more effectively Wade examines the structure of incentives and supervision in the two national departments, and the role of the patroller who links the bureaucracy with the clients. The Indian patroller is a full-time employee who is appointed to the area by the engineer in charge and then rotated every six years and not allowed to work in his native area. The thrust of these rules is to minimise contact with the locality and maximise that with the bureaucracy.

The Korean incentives are the reverse; the patroller is a part- time employee who must farm in the area in which he works. He is appointed by the headman and must be re-nominated every year. The supervision arrangements in the two departments differ as dramatically as the incentive structure. In the Indian department, supervision proceeds up to the central level in a strict geographical hierarchy.

By contrast, the supervision arrangements in Korea operate in a much flatter hierarchy; the supervisor travels up and down the command area, observing the operations and stopping to chat with the patroller and farmers. Wade argues that the essential difference in the performance of the two departments is that 'the Korean type of irrigation structure itself maximises the spillover of community social capital into the irrigation organisation and then generates more inside the organisation, while the Indian structure minimises it'. An attempt to build institutional capacity would most probably miss the mark if it did not recognise the importance of these wider relations in organisational behaviour.

The Indian system, with a high cost of moving information, a single tall hierarchy, and no downward accountability, permits opportunistic behaviour between the department and their clients. These case studies are interesting for the case of watershed management because they illustrate how bureaucratic structures can influence the behaviour of officials and also, how the operation of these structures cannot be understood without considering the effects that they have on the informal relations between the bureaucracies at the 'street level' and their clients. The Guidelines have made a commitment to changing both bureaucratic structures and behaviour. The emphasis has so far been based more on the construction of complementary relations than on increasing the embeddedness of the relations between state and society.

ENDOWMENTS' AND THE POTENTIAL OF INSTITUTIONAL CONSTRUCTION'

The notion of complementarity and embeddedness are useful in an understanding of different elements in state-society synergy. However they are descriptive and leave the question of 'how to get there' unanswered. The concept of endowments and constructability are fundamental to answering the question of how to create state-society relations. Does the creation of such relations depend on pre-existing endowments of social capital? Or can the application of 'soft technologies' such as capacity building of various kinds construct synergy over relatively short periods of time?

These are some of the questions that arise when considering the institutional alliances which the Guidelines are based on. Communities are usually not homogenous, governments and NGOs compete, and governments are not accountable to people.

Can these relations be reversed within the cycle of a watershed project? At least reversed enough so that the benefits of watershed management are sustainable? 'What of the empirical evidence of the importance of endowments of social capital in the three parts of watershed projects: civil society; the public administration and the political system?' Putnam (1993) has stressed that good governance is created by social capital that is accumulated over long periods of time in civil society. The imputed scenario for watershed development is the proliferation of CBOs of different types who build networks of local trust and reciprocity.

These networks will in time create good and accountable government from the bottom up. But how long will this take? The case studies in World Development find to the contrary that whilst micro-level social capital, such as exists between homogenous communities is positive, it is not a key constraining factor in successful state-society synergy. '

The limits seem to be set less by the initial density of trust and ties at the micro-level and more by the difficulties involved in "scaling-up" micro-level social capital to generate solidary ties and social action on a scale that is politically and economically efficacious'. An exploration of irrigation systems in Taiwan collective action in Mexico and, most relevantly, development in Kerala all find that the stock of social capital were not the main reasons for state-society synergy. Heller (1996) argues that Kerala's success in community mobilisation is not a consequence of pre-existing social ties. Traditional parochial loyalties, underlined by caste, would have resulted only in the same type of vertical linkages that are a problem in the rest of India.

Instead, success lay in relating these ties to a more universalistic set of identities, and scaling-up personal and community ties into encompassing forms of organisation that can effectively pursue developmental goals. This conclusion has very positive implications for the operation of the Guidelines.

What endowments of social capital are necessary in public administration and what shape should state-society relations take? There are two competing perspectives on this question: one that emphasises the importance of Weberian bureaucracies in making sure that embeddedness does not become a cause of clientelism. The other perspective emphasises the importance of decentralising the bureaucracy and sensitising it to in-puts from the 'bottom-up'. Both perspectives are represented in discussions over the role of the bureaucracy in India and both are evident in the Guidelines. Whilst there are hopes for a 'quiet revolution' in India through a reorientation of the bureaucracy to the grass-roots, this is not yet represented through a change in the structure.

The case studies in World Development, especially those of Taiwan, Korea and Kerala, suggest that a tightly controlled hierarchical bureaucratic structure does not negate the construction of state- society synergy. In fact, in Taiwan the structure of the irrigation department gives farmers the confidence that water will be delivered and increases the incentive for cooperation at the local level.

The tight bureaucratic structure does not interfere at the grassroots, leaving local officials and people to work out their problems. What is critical in the operation of the bureaucracy is the structure of incentives, authority and responsibilities. In Kerala, land reform and the identification of party and state officials with their constituents has been a critical part of the construction of state- society synergy. In Taiwan, authority at the top is closely linked to responsibility, and incentives at the bottom are closely linked to an accountable performance.

The public administration in India, is weak in both areas. What of political interests and their effect on endowments of social capital? This is the third influence in the operation of the Guidelines but it is often relegated to the background in discussions. Further much of the discussions of state-society synergy, recognise conflict but then goes on to ignore the operation of this in practice. Most of the studies implicitly assume that communities are relatively homogenous. They then extend that assumption to the public sector and assume that this sector has shared interests with its constituents.

The creation of state-society synergy then becomes a question of overcoming collective action problems. However what if a community is divided by conflicts created by property rights and caste, and the public sector is not accountable to its clients? The question that then arises is whether political competition is a useful way to resolve these conflicts. 'Political regimes no less than bureaucratic structures condition the possibility of synergy and social capital formation'.

The World Development case studies generally find that political competition has a positive influence on the creation of social capital. The case study of Kerala illustrates how political competition has sustained the commitment of parties to mobilisation and capacity building, amongst weaker groups.

Competition can help prevent the ability of individual élites to interfere in development projects and make projects accountable to local people. However there are two preconditions for the effectiveness of political competition in creating state-society synergy. First the competition must follow mutually accepted ground rules, and power groups must not be allowed to interfere with violence or coercion. Second, a technical issue, there must be an efficient administration which can deliver the goods and services offered by competition.

THE PROCESS OF INSTITUTIONAL TRANSITION

State 'failure' in development planning is a widely acknowledged fact and has led to the recent moves to reform the state and to involve other institutions. Complementarity in state-society relations is generally acknowledged as a critical component of successful development programs. The Guidelines are a recent and innovative attempt to build institutions according to a stylised ideal of state-society interaction for development planning. There are two critical questions that an examination of the system in practice raises. First, how can we explain differences in institutional performance? Secondly, having identified the desired institutional combination, why do we not establish these?

If rigid bureaucratic structures with flexible local officers are the ideal then why not change all line departments to operate in this fashion? If political competition is positive for breaking traditional caste and hierarchical alliances then why not allow such competition to flourish? The Kerala design is widely acknowledged to have enabled development: so why not replicate it throughout India?

On the first question, Khan (1996) has argued convincingly that understanding why similar institutions perform differently depending on the context requires an understanding of the political settlement. The political settlement is a balance of power: the balance of power determines what is demanded, by whom, and on what terms. It also determines the feasible responses of the state to such demands, given the structure of supply'.

An understanding of the political settlement is critical in understanding how institutions work. Why, when the Guidelines are so uniform, is their implementation so different in Karnataka, Andhra Pradesh and West Bengal? 'One consequence of recognising differences in the political settlement is that it is possible to explain why performance rankings of institutions in one political settlement may not be transportable to another'. Andhra Pradesh has a relatively accountable bureaucracy, but it would not be possible to reconstruct this in Orissa without an understanding of the political settlement within that state. Similarly, it is possible to decentralise a political system without changing property rights or the political settlement through which it operates. A comparative analysis is therefore useful only if these differences in the balance of power are taken into account. Which leads to the second related question; why are successful models so difficult to replicate?

The notion of transition costs suggested by Khan is useful in providing a framework within which to ask this question. He argues that 'most important institutional changes are politically resisted by the losers the intensity and extent of resistance is the real 'cost of change' faced by its initiators, namely, the transition cost'. This transition cost is one of the main reasons that institutions do not simply change in the direction of increasing efficiency, for as Khan notes, even a predatory state could benefit from growth.

So, it may be more efficient for the entire system if development projects reached their clients and stimulated growth, but such a transition faces political obstacles. This notion of the transition, and the costs of transition, was one often remarked on by the respondents in the study. 'The government will not like to give up its power.

The Collector now is like a Nawab during the colonial times. He is a very important man and everyone gives him respect.' These transitions costs were most often mentioned in relation to Panchayati Raj: 'It will take time because there are vested interests who will lose if PRIs become powerful' was one explanation, 'it is important not to be impatient, Panchayati Raj is part of a long-term programme. There are powerful interests against it at the village level. We need a long-term programme to change these interests'.

The transition period, for PRIs, was most commonly placed at around ten years. 'Everyone says that Panchayati Raj is a good thing, but this is only lip-service. The bureaucrats and Members of Legislative Assemblies (MLAs) are against it'. The following comment is also representative: 'it will take a long time to change which is why we have to start working with PRIs right away'.

CONFLICT AND COLLABORATION IN THE IMPLEMENTATION OF THE GUIDELINES

What are the transition costs in the manifestation of the Guidelines? What is the political settlement that will influence the way in which they are implemented? The Guidelines will have to operate with already existing institutional alliances and conflicts. For example: alliances and conflicts between sectors (government, NGO) within the government (centre, state and district); and between PRIs, governments and NGOs. This section will briefly consider some likely trends in the relations between them based both on the literature and on the fieldwork in Karnataka and Andhra Pradesh.

Party Political Conflict

Panchayati Raj has been used as a political tool, in various ways, by all of the governing parties. This is not inherently bad, but very few have delivered on promises to establish an empowered Panchayati Raj system. The congress government oversaw the establishment of PRIs in the 1960s but by working around them and suspending elections, rendered them virtually useless. Instead PRIs became a means through which the congress government could gather vote banks in exchange for the selective distribution of some development projects. The major movements to re-establish functioning PRIs in the last decades have all come from state governments, often as part of their drive to gain autonomy from the centre.

The governments of West Bengal, Karnataka and Andhra Pradesh have all at various times included PRIs as part of their election manifesto. This has often been established in the face of enormous opposition from the central government: the Minister for Panchayati Raj in Karnataka only managed to get the GOI to pass the Act after threatening to fast unto death.

PRIs established by a given government are often perceived to be a threat by succeeding governments who perceive them to be the local power bases of that regime. Thus the first thing that congress did on being re-elected in Karnataka was to roll back many of the powers that the Janata Dal had given to PRIs. Panchayati Raj will always be affected by the agenda of the Party in power at both the centre and state level. This is inevitable, and in itself not a problem.

The problem lies:

- In the fact that most of the parties are reluctant to decentralise real power down to PRIs.
- That they are able to give and take power away from PRIs, more or less as they wish.

However, PRIs are increasingly able to exert pressure on the government, whatever the party, to devolve power. 'So far as political parties are concerned, whatever may be their reservations, and some reservations still continue, they should recognise that in recent years there has been a great deal of public debate and discussion and indeed demand for decentralisation. And I think what has lent further power to this demand is our accumulated experience of development in this country over the past four decades. The crux of this lesson is that development cannot acquire dynamism, equity, speed and fuller mobilisation of resources and their efficient use without the aid of decentralisation'.

The 73rd Constitutional Amendments are evidence of this recognition that some power will inevitably have to be decentralised. A retired Indian Administrative Service (IAS) man commented 'the government feels as if it is holding a tiger by the tail. They know that in the long run they cannot hold a tiger, but they are scared that if they let go it will bite them'.

ADMINISTRATIVE AND POLITICAL CONFLICT OVER DECENTRALISATION

The single biggest obstacle in the way of Panchayati Raj is political and administrative conflict against decentralization. This is common to all political parties, although less of an issue with cadre based parties like the Communist Party of India (Marxist) in West Bengal. It is also not a single 'organized' obstacle against decentralization, but a complex web of shifting alliances the operation of which is a much-debated subject. This section can therefore only present a stylized version of some of these alliances and some widely accepted arguments about how they operate. One of these general trends is that the central government, until recently mostly congress but this is true for all Parties, has been opposed to decentralization.

Many of the subjects that were on the state list have been shifted onto either the concurrent or the central list. Therefore in practice the centre has taken functions, powers and authority that are well beyond what had been allotted to them in the constitution. The enormous growth of centrally sponsored poverty alleviation schemes on subjects that are on the state-list are a good example.

Many observers argue that the first necessary step in decentralization, either for Panchayati Raj or development programs, is the restoration of powers to the state. The centre prefers to allocate them directly to the District Rural Development Agencies, which are registered societies established especially for this purpose. This conflict between the centre and the state is still apparent and is a major conflict in the implementation of both Panchayati Raj and the Guidelines.

Respondents at MRAE in New Delhi said that the reason for sending funds directly down to the DRDA, is that if they are sent to the state, they will take a long time to reach the district, and may be temporarily diverted for other purposes. In fact at the centre many expressed a preference to by-pass even DRDAs and Zilla Parishads and channel funds directly to Gram Panchayats and Watershed Committees. This proposal was incorporated into the approach paper of the ninth five year plan. A respondent in the Department for Rural Development in Delhi said that strengthening the Gram Panchayat was imperative. 'Let the Zilla Parishad and the DRDA fight it out. The ZPs can take it, they are not innocent fellows. We should target the Gram Panchayat directly'.

So what is the problem with this? I asked a retired IAS respondent and he commented 'it is like a vote of no confidence against the state governments'. Nobody denied that the state governments are slow to release money, but they all argued that bypassing the state was not the solution. Most commentators would argue that the centre does not have the necessary information to make decisions about what is appropriate at the village level. Chief Minister Naidu of Andhra Pradesh criticised this process: 'even what is to be done at the village level is being decided in Delhi'.

Further, the funds are allocated through political filters and have political favours that have to be repaid in return. A retired IAS respondent: 'they are sending the funds straight to the district level because they want people to think that it is only the central government that can help. It is a populist move'.

One of the arguments made at the Centre for Direct Intervention at the district level is that state governments themselves centralise power. This is also generally true. One of the biggest sources of opposition to Panchayati Raj has come from the state level and from MLAs in particular. This is true even of states like Karnataka which have introduced fairly progressive Panchayati Raj legislation. Today the biggest problem in decentralization is the opposition of MLAs. In particular they are reluctant to give powers to the Taluk Panchayats, because that is their constituency.

This is a generally accepted fact, repeated often in both Karnataka and Andhra Pradesh. The MLAs have a considerable amount to lose from a fully functioning Panchayati Raj system, especially if the Taluk Panchayats are fully empowered.

The Taluk is their constituency, and it is there that they harness votes and deliver favours. 'Even the Janata Dal MLAs were not fully convinced of the rationale for the erosion of their power and status. How can lesser mortals – the small village leaders – usurp their (the MLA's) powers "to hold all that they surveyed", the latter would ask'. The MLAs have fought hard to prevent any real devolution of power. The most evident manifestation of their control is that they are members of the Taluka Panchayats of their constituencies and therefore have a direct influence on developments there. In Karnataka the MLAs have an 'MLA development program' which directly competes with the development programs of the Zilla Parishad and is a clear avoidance of the spirit behind Panchayati Raj.

Several respondents at the workshop on watershed development in New Delhi mentioned conflicts between the MLAs' contractors and the Zilla Parishad contractors for tenders. The centralising tendencies, and the conflict between the state administration and PRIs, continue at the district level and below. Within the Panchayati Raj system the centralising tendencies work against the transfer of powers from the Zilla Parishad to the Taluka Panchayats and the Gram Panchayats. Within the state administration the tendency for centralisation within line departments, for the long line of command and upward accountability, and for Collector Raj, is well-known. The struggle between the state administration and the PRIs for control is also most apparent at this level, and there was plenty of anecdotal information from the respondents about this conflict.

It is at the district level and below that many of the votes are gathered, that funds are diverted, and that promises are made. In the context of watershed management this is especially apparent in the frequent vague mentions of Contractor Raj, of projects being delivered to specific blocks, and of favoured appointments of PIAs, and WCs. These jobs, positions and allocations, have a price on them, as was frequently pointed out in Karnataka.

It is at the district level and below that many of the costs of transition will be determined. Khan has noted that it is often assumed that the political settlement that underlies the operation of the state is patrimonial. That is, the state is the 'patron' that distributes benefits to its clients for a share of rents or political support. In fact the evidence suggests that this vision does not depict the reality of the operational constraints on the state. Khan proposes an alternative depiction – the clientelist political settlement – that recognises that states often lack legitimacy.

The main characteristic of the clientelist political settlement is that 'the property rights defined by the state are weakly defined and are contested by well-organised social groups who are able to challenge the rights being enforced by the state'. This alternative conceptualisation of the dynamics underlying transactions between the state and civil society has several implications for a consideration of transition costs. Because existing rights are likely to have been negotiated politically in the past, it will be difficult for a weak state to withdraw and transfer these, because the political costs will be too steep. The state may thus prefer to create new rights than transfer existing ones.

RELATIONS BETWEEN PRIS AND THE STATE ADMINISTRATION

Many of these structural conflicts over power and resources are manifested in differences over work patterns, incentive structures, and work cultures between PRIs, NGOs and the state administration. Some of these differences are themselves structural and some are more behavioural. There has been plenty of work done recently on differences between the state administration and NGOs, as well as plenty of good work on how some of the more superficial differences can be resolved. Many initiatives to resolve these conflicts have been relatively successful; the partnership between the government and NGOs for watershed development is one example of this. There is still plenty of conflict; the government can be reluctant to make NGOs PIAs, and NGOs are often highly critical of the work-culture of line departments.

However much progress has been made and NGOs are now involved not only in implementation but also increasingly in formulating policy and planning. A lot of attention, much of it supported by donors, has been placed on establishing areas of complementarity between the government and NGOs and building on these.

The relations between the state administration and PRIs are more complex and there has so far been less room for cooperation. On paper the complementarity of these two systems is obvious and an integral part of any democratic system. However the bureaucracy has since Independence been given such a central role that it is now unwilling to consider itself as a tool, a merely administrative arm, of elected representatives. This fact was clearly established and confirmed by all the respondents.

The only difference in opinion was over whether this opinion had any merit or not. In the opinion of a personal assistant at the Department for Rural Development and Panchayati Raj 'the Adhyaksha and the members of the Zilla Parishads are only elected fellows. They don't understand anything about how to do development work'. Another respondent commented 'the Collector will not like to have an elected fellow telling him what to do. He is used to being King in his district'. The Ashoka Mehta Committee on Panchayati Raj in 1977 found the conflict between the administration and the PRIs to be the single biggest cause for the failure of PRIs.

'Bureaucracy has probably its own role in disassociating the PRIs from the development process...The system of line hierarchy would find favour with them as an organisational principle. The officers would feel that they are primarily accountable for results and financial proprieties to the state governments. The officials know no better than to trust their own fraternity.

They would, on the one hand therefore be averse to PRIs being entrusted with additional functions and on the other would not easily get adjusted to working under the supervision of elected representatives'. The conflict between PRIs and the state administration, and for that matter NGOs, is also one of class. The PRIs may be dominated by the local elite, but they are a local élite. Government officials, especially IAS, and NGOs, often have whatever their differences, a common class background.

The informal links between the IAS and NGOs in particular are manifold, with an increasing slippage from one system to another, English as a common language, and a common middle-class, often urban, background. There is considerable room for embeddedness in the construction of positive synergy between the IAS and NGOs. A respondent from an NGO in Bangalore commented 'now we have close working relations, and we are realising that our differences are often just ones of perception. Many of the IAS officers are becoming more aware of participatory approaches. There are several officers who come around to talk and we give them informal help if they are having problems with some of their work'.

The potential for such relations is far less between the IAS and NGOs on one side and PRIs on the other. A former Secretary of Agriculture commented 'the IAS is an élite system and we are totally different from those who are elected to PRIs. I think there is no difference between an Adhyaksha and a MLA. But most people do not agree and do not give much respect to the elected members in PRIs'. Most of the IAS respondents however, and this is of course a subjective comment, perceived the PRI has populated by backward, short-sighted and feudal people.

This is also true of many of the NGO respondents. Mathew has commented on the preference of many NGOs for working with the government: 'they prefer to deal with the district collectors and senior government officials rather than with the unsophisticated, rugged villagers, that is to say, these middle-class, English- educated social workers and activists are more at home with their own class in the bureaucracy than with the villagers'.

The same argument has been made about MLAs whose alliance with the state administration has so far been stronger and more enduring than that with other elected representatives in PRIs. I have so far concentrated more on areas of conflict than on collaboration between the state administration and PRIs. Perhaps this is because their record of constructive and collaborative work has been short.

There has however been some positive experience of complementarity in work between the PRIs and the bureaucracy, particularly in Karnataka after the 1987 Panchayati Raj legislation was passed. Instead, resources should be made available to Panchayati Raj Institutions with mechanisms established for the monitoring of resources and programmes. However the Karnataka experiment was relatively successful, with a clear demarcation of power, incentives, and suitable promotional avenues for senior functionaries working underneath PRIs.

Although these are still volatile, at the moment their effective resolution is despite of rather than because of legislative support from the top. Finally, the judiciary has often supported Panchayati Raj legislation against the arbitrary decisions taken by state legislatures. For example, the Karnataka high court responded to a petition presented by fifteen citizens that their Zilla Parishad elections had been delayed and ordered that elections be held.

There have been similar actions and responses in Orissa and in Uttar Pradesh. *NGOs and PRIs* What role will NGOs play if the Panchayat system starts to work and how do NGOs feel about the Panchayati Raj system? The answer from one respondent in ActionAid is representative: 'frankly, NGOs feel threatened by Panchayati Raj Institutions'. The role and scope of NGO involvement in development projects, development programs and now even development policy formulation has grown enormously in the last two decades.

Their role has in part been facilitated by the lack of a viable grassroots system for promoting local development. It has also been facilitated by the increasing allocation of funds available for NGOs. 'Over Rs.2000 crores comes from abroad every year through the Foreign Contributions Act and the so called registered voluntary societies receive another Rs. 10.000 crores from foreign funds'.

The Department of Wasteland Development (DWD) within MRAE administers a grant-in-aid scheme in which funds are allocated directly to NGOs specifically for wasteland and Micro-watershed development. The potential conflict of interest between NGOs and PRIs revolves both around the role that they would play and the funds that they would access. One of the main reasons that PRIs failed to operate in the past is because of the non-availability of financial resources and the absence of clearly defined financial powers.

The State Finance Commission now has the mandate to establish the financial powers of the PRIs, both through allocating them funds and specifying areas in which they can raise revenue. Suggestions have also been made that NGOs should not receive funds from either the government or foreign agencies.

'Instead Panchayati Raj Institutions may be strengthened through opening of new resources for their income with critical backing from resource agencies established for the rigorous monitoring and management of programmes and resources'. However there is also room for collaboration between PRIs and NGOs. It will take some time before PRIs are operating effectively and it is quite likely that PRIs in the immediate future will be run by local élites. For that matter, anyone who is elected will feel the pull of a power structure where the incentives lie upwards.

This pull will only stop when the Gram Sabha is working effectively and when elected representatives are locally accountable. NGOs have much potential to fulfil this role. Many NGOs have over the last decade been specialising in 'capacity building' skills, which provides an ideal basis for informing people of their rights and encouraging them to act collectively to claim them. A close consideration of the subjects in the XIth Schedule which are under the jurisdiction of PRIs, and those issues in the seventh plan identified as being the role of NGOs, reveals that there is room for cooperation. Most of the XIth Schedule concentrates on subjects that should be developed by the PRI (such as: watersheds, schools and drinking water).

The description of the role of NGOs in the seventh plan concentrates on facilitation: 'to supplement the government efforts…'to disseminate information' … 'to de-mystify technology'. Whilst there are certainly areas of conflict, there is also a lot of room for collaboration. Many NGOs have already taken this initiative and have started working directly with Gram Panchayats to strengthen their capacity. Others are encouraging the Self-help Groups (SHGs) that they have formed to run for elections to the Gram Panchayat or to place pressure on the Gram Panchayat to listen to their needs.

Development planning and development in India are currently undoubtedly in a period of transition. There is a widespread recognition that the state has 'failed' in development planning. The general solution has been both to decentralize and to involve a range of civil society institutions and develop areas of complementarily between them. The Guidelines for watershed development are a representative example of a development programme that has tried to institutionalize this approach. However these Guidelines are built on pre-existing alliances between different institutions, and between levels within the same sector or institutions. An understanding of the operation of the system, in particular of the links between PRIs and WIs, needs to examine how these systems operate and likely points of conflict and collaboration.

The stylized account I have presented of these points of conflict and collaboration is almost entirely based on comments that were made during the fieldwork. An analysis of these areas of conflict, and how they can be resolved, will depend on the particular view of the political economy of India taken by the observer. The operational implications of viewing the state as an agent of the dominant ruling classes, as an impartial agent with no room for autonomous action and as a predatory state driven by rent-seeking impulses are entirely different. However, whatever the political spectacles, an understanding of how the Guidelines operate in different states and the links which exist between PRIs and WIs, is only possible through an understanding of the balance of power between the groups involved. Further, building institutional alliances, and changing structures is usually not a technical, but a political exercise.

This is because all such changes have transition costs and it is important to be explicit about the values and notions of justice on which particular choices are made. Khan notes the importance of making these costs explicit and avoiding a false sense of objectivity in deciding between alternate institutional arrangements. 'It is worth remembering that a 'mistake' in the assessment of transition costs involved in implementing particular programmes of institutional change can ultimately result in civil war and a large scale loss of lives'. Most of the speculations and comments about transition costs mentioned above, and discussed in the interviews, were about the 'modern sector' of India.

Those who are in formal institutions, already have access to resources, and are reluctant to part with them. Most of the arguments this sector makes about desired institutional systems are made in the name of those who are not in these institutions.

Many of the IAS officials commented that the people are not ready for Panchayati Raj because, basically, the transition costs, will be too high. This is undoubtedly a real danger: The costs of challenging élite power structures in a village in Bihar are very real, for example. Others argued that the people had suffered enough through Collector Raj. As a proponent of Panchayati Raj comments, the central question in development planning continues to be what it was 100 years ago: 'that is, what place should be given to the Collector rather than what place should be given to the community'. Both sides concede that ultimately the effective functioning of both the administrative and the political system depends on a local community, whether the Gram Sabha or a Watershed Committee, which can demand accountability and ensure equitable distribution.

COLLECTIVE ACTION AND EMPOWERMENT

The arguments for decentralisation and different types of institutional collaboration examined in the previous sections are all based on assumptions about what induces local collective action. The notion of participatory development which underpins the Guidelines and Panchayati Raj are both justified on particular visions of how they can induce local development. Despite these claims, the evidence on the operation of these systems is always the most vague and bare at the grassroots. Within the Panchayati Raj system, there is more evidence of how it operates at the Gram Panchayat level upwards than on the effect it had on the Gram Sabha.

Within the Guidelines, the discussion and evaluation has concentrated on the performance of the various actors at the intermediate level. The information on how watershed projects have affected the poor is vague: 'élites still dominate', or 'the main benefits are wage employment' are fairly common statements.

Little is known of the processes through which local people become empowered in either system. This section will consider explicitly the notion of local collective action and of community empowerment implicit in Panchayati Raj and in the Guidelines. I will argue that they are very different, but that recognising their difference can be a means through been on which to build on their strengths. Some of the recent work that has deconstructed different types of social capital is very useful for this purpose. This section will briefly consider the potential of both Panchayati Raj and participatory development in externally funded programs.

THE COMMUNITY AND DEVELOPMENT IN PANCHAYATI RAJ AND THE GUIDELINES

Panchayati Raj and the Guidelines differ in their interpretation of how to promote collective action, and how to relate this to community development. Both systems have different visions of how the community operates and how to build social capital. Social capital has been defined as 'features of social organisation, such as networks, norms and trust, that facilitate cooperation and coordination for mutual benefit'. An exploration of the debate about how to build social capital is useful because it asks the question: is social capital a necessary basis for effective local government? This question lies at the heart of much of the debate on decentralisation and institutional change explored in the last two sections. In the context of the comparison between PRIs and WIs and the linkages that exist between them the critical question is:

- Can Watershed Institutions provide a basis for the establishment of social capital, and if so, can?this be institutionalised and scaled-up?
- Can Panchayati Raj Institutions provide a basis for the establishment of social capital, and if so,?can this be institutionalised and scaled-up?
- How do the two systems compare in their capacity to build social capital?

The revived interest in social capital grew out of some comparative empirical work on the social, political and economic development in north and south Italy. Putnam argues that the most important determinant of the performance of democratic local governments, and of differing levels of socio-economic development is 'civic tradition' and the level of 'civic involvement'. Strong interpersonal ties (like kinship) are less important than weak horizontal ties (like membership in an association, which could be watershed management) in sustaining collective action and building capital. Putnam argues that in southern Italy such horizontal ties are weak and institutions are locked into a vicious spiral that reflects this pervasive mistrust of each other.

The north, by contrast, has experienced a long history in which networks have been established and social capital has been developed. This has led to positive cycles of cooperation and 'norms of generalised reciprocity' in which cooperation is the first tactic. The central variable here is trust, which comes through repeated interaction and sanctions for defection.

The implications for building social capital in the context of an Indian village community is that associations, such as Watershed Associations, and all of the other user groups which are being formed, slowly establish horizontal ties of collective action. These cut across kinship ties and develop norms of cooperative behaviour through repeated interaction and sanctions.

These in turn develop the social basis and demand for effective local government. So, the growth of 'associational life', through participatory development, will build the social capital on which effective government, Panchayati Raj, has to be based. This argument is convincing and in the context of PRIs and WIs has many adherents.

However there are some difficult questions about how this process of 'associational life' starts and how to scale it up. Putnam has noted that 'most institutional history moves very slowly'. Does it mean heavy NGO involvement in every village? And for how long? How does one break the vicious circle of 'pervasive mistrust' and start building 'norms of generalised reciprocity?' Some of the responses to Putnam's work are relevant for a consideration of the social dynamics that exist and have the potential to form in PRIs and WIs.

First, Putnam argues that social capital is the property of a group, but that everyone can become part of this group, and that it has no negative externalities. But groups, by definition, are exclusive. Putnam does not account for the possibility that they may not necessarily co-exist peacefully with other groups, and that they may have negative externalities.

The elections to Watershed Committees, for example, are now often being fought on party political lines. In many villages Watershed Committees become powerful pressure groups, and have the power to take decisions, such as closing the commons, which may go against the needs of another group. It is therefore not necessarily true that the more 'associational life' the better, or that the social capital that is created by these groups is there for all.

In fact many observers have noted that there is a 'civil society grab' with associations at both the village and intermediary level, taking advantage of the opportunities now being provided by development programs, and crowding out others. Secondly, and related to the above, there is no political variable in the explanations Putnam provides for the construction of social capital through 'associational linkages'. How do these tightly knit networks of civic engagement lead to more effective government? What mechanisms would be used to make the government more responsive to their needs?

For example; how would a Watershed Committee influence the MRAE to provide more or less of a specific input? And, related to the above, what guarantee is there that the claims which are made are representative of the wider society. The answer to these questions depends critically on the wider political setting and the means through which groups can press demands.

Thirdly, there is significant evidence that participatory development and the establishment of civil society links do not necessarily make for better government. Tendler (1996) has argued that there is at least as much evidence that the causal relationship goes the other way. That is; basic structures of local government are created in which 'associational life' and social capital can grow. Some critics have taken another look at the empirical work on north and south Italy and argued that Putnam has missed an important element in the cause and effect cycle of social capital.

The underdevelopment of southern Italy can be explained by the fact that it was semi-colonized by the north and that no room was provided by the state for local government. Conversely, the performance of regional governments in the north today may also be explained by the legacy of progressive political parties operating in the area in the nineteenth century. The combined directions of these criticisms of the notion of social capital formation cannot be considered without the political context, or as an alternative to local government.

The combined articles from World Development all argue that a minimal stock of social capital, in the form of kin relations or cooperative work relations, will exist in almost any community. There is nothing wrong in building on these relations through forming associations, or SHGs. What makes the critical difference is how this social capital is scaled-up and there is considerable evidence that associations themselves are not sufficient for this purpose. This criticism is very relevant for a consideration of Watershed Institutions, which have been established to build local capacity and social capital.

The question, which I will leave hanging for a moment, is whether WIs, provide an avenue through which social capital can be scaled up, and if not, whether this role can be taken by PRIs. This exercise of distinguishing different types of social capital is useful because of the current enthusiasm for building social capital through 'local institutions'. Local institutions can build social capital through information sharing (credit schemes); coordination of activities (water user groups) and collective decision-making.

All of these different types of social capital are themselves affected by the 'macro-level social capital' in the institutional framework of society as a whole. But there is very little exploration of how different institutions interact. This is true of the Guidelines, where the Watershed Committees and Gram Panchayats are mentioned interchangeably in the text, although their functions and potentials are very different. There is also very little exploration of the political dynamics of the system. As Harris (1995) notes, the World Bank Report recognises that local institutions are more effective 'when the local distribution of assets is more equal', and that 'the application of social capital in development is not a distribution neutral process', and that by themselves, without other resources, 'these institutions may not make much difference'.

Yet the recommendation remains that donors and the government should promote 'desirable forms of social capital' as if that makes it any clearer. This broad sweep across the notion of social capital means that the term has attracted a diverse constituency. This was apparent in the discussions with the respondents in my interviews. Some respondents saw WIs as part of an effective delivery system and put primary emphasis on effective watershed management, and the economic benefits generated, to provide a base for further development. Others said that whilst economic benefits are critical, the formation of WIs is a vital end in itself as a contribution towards local empowerment.

There is also a difference between those who consider PRIs as a part of the delivery system, and those who see their inclusion in the Guidelines as an opportunity to strengthen the democratic system. There was also a big difference between those who saw WIs, along with other SHGs as a necessary foundation for PRIs, and those who considered that PRIs were necessary to scale-up social capital.

COLLECTIVE ACTION AND EMPOWERMENT IN PANCHAYATI RAJ

There was a general recognition amongst all of the respondents, and in the literature; that the effectiveness of Panchayati Raj ultimately depends on how well the Gram Sabha works. The whole democratic principle is that the relevance and the accountability of the system will be determined from the bottom. The Panchayati Raj system depends on the operation of the Gram Sabha. If the Gram Sabha is working then automatically all the other levels will work also. There was nobody who disputed the ultimate importance of having a local democratic system. The arguments for and against Panchayati Raj are based mainly on different opinions of whether the Gram Sabha is ready to play such a role and if not, why not.

Similarly, there is a consensus that part of the reason that Panchayati Raj failed in the past was because it was opposed by vested interests, deprived of funds, political support, or executive autonomy. The debate centres on what would have happened if it had external support and all the resources necessary to function as a system for local self-government.

The basic issue underlying all of these debates is whether 'the people' are ready for a fully operating Panchayati Raj system. In all of my interviews about the role that PRIs should play in watershed management, this was the basic question to which we returned. The most well known early proponent of social change first, PRI later, was Dr. Ambedkar. In his famous argument against giving Panchayats power after Independence he asked 'what is the village but a sink of localism, a den of ignorance, narrow-mindedness and communalism?'

Many still argue that it is imperative that democratic decentralisation should be preceded by land reform and literacy campaigns. Without this, democratic decentralisation will simply end up as a means to decentralise responsibility without authority. These fears are not groundless, and the consequences of local institutions vested with power is that the élite often do not merely take control, but violently enforce it, particularly when it is challenged.

This has led observers to argue that expecting the disadvantaged to take such a risk, even if they have been elected through reservations, is optimistic without providing them with support. In Madhya Pradesh, which was the first state to hold elections after the Amendment, there was a report of a woman in the Gram Panchayat being raped, another being publicly stripped, and a scheduled caste man being tortured and beaten. In Tumkur, in the Taluk I visited a scheduled caste man who had been made to eat human excreta.

How can PRIs be anything but a farce in villages where the system basically remains feudal? There are several arguments made for the immediate transferral to PRIs of the powers with which they have been vested by the Constitutional Amendment. The first is that the state administration has had 50 years to promote community development and has not been able to achieve this.

The second is that 'a Panchayat is a microcosm of the society of which the village forms a part, and should be part of the process through which local structures of inequality are changed. Further, democracy requires practice. A current proponent of Panchayati Raj said in response to the quote by Ambedkar given above: 'what instruments did Ambedkar leave behind or has anybody devised in the past 40 years by which the village he described could be or has been changed? In forty years at least in 1 lakh villages out of 5 lakhs a change would have taken place if they had periodic elections'.

Hirschman's (1970) formulation of exit, voice and loyalty are useful for a consideration of the dynamics of change operating in villages today. The first option, exit, breaking all relations, is preferred when the outcomes of continued interaction are all negative, although exit may have high costs (defamation, loss of livelihood, etc). Loyalty represents a situation where the interests of all members are integrated and the rules are not contested. Loyalty and rudimentary forms of cooperation have been sustained over the years by traditional authority structures. Although the behavioural norm was loyalty, these were often based in unequal patron-client relations.

Voice stands for a bargaining situation, in which the individual or the group does not exit, but also does not remain loyal, and instead decides to change the terms on which they are integrated into the community. Many of the respondents in this study argued that the Indian village community is changing rapidly. A large part of the reason for this is the advent of participatory politics that is eroding traditional authority structures in myriad ways.

Bardhan (1993) argues that whatever the final effect, they improve the options that common people have for 'voice', 'exit' and the construction of new social relations. He notes that there are 'appeals to supralocal authorities for conflict resolution and arbitration' and that 'with increased social and political awareness villagers are, however, increasingly able to differentiate between domains of relationships which formerly used to be lumped together in a dense all-encompassing network'.

For example, several observers argued that Radegaon Sidhi was not replicable because people would no longer consent to hardline rules from an upper caste individual whether it was for their benefit or not. In the Gram Panchayat I visited, several of the upper caste members commented on 'the times' and how they are finding it very difficult to enforce their decisions.

As one respondent mentioned 'the élite know that if they are not careful they will kill the golden goose'. The argument that democracy needs practice, and that PRIs should be part of the process of development and change, are at the heart of the case made for empowerment of PRIs. There is considerable evidence from the first phase of Panchayati Raj in the 1960s that some places were beginning to benefit from the system before elections were stopped and it was derailed.

There is evidence that local people understood the Panchayati Raj system and were aware of the significance of the transfer of power from the Block Development Officer into their hands. Further, there existed the rudimentary beginning of local development plans. The report found that conferring power to people's representatives had improved the attendance of teachers in primary schools, while block administration had become more responsive.

It is widely accepted that the effectiveness of local government is one of the explanations for the development of Kerala. Observers also point to the past experience of Panchayati Raj in Andhra Pradesh and Karnataka has an important part of the explanation for the political awareness in those states.

Mathew notes of a meeting of Zilla Parishads and Mandal

Parishads in Karnataka that 'their leadership, boldness and sense of achievement were especially noteworthy. There was a general agreement that in the last five years resource mobilisation had been excellent and sources that had been previously untapped have been tapped. Works, which had not been taken up for thirty years, had been implemented during the term in office of these leaders.

A woman leader from Neelamangala had recounted that till the new Panchayati Raj came into existence she was the only woman in that area who used to come out in public. But after five years of Panchayati Raj the situation has changed dramatically; many women now participated in the affairs of the village. The evidence of the breakdown in caste based alliances in Karnataka, and the empowerment of women through Panchayati Raj, is overwhelming.

After the Panchayati Raj Act was passed in Karnataka in 1987 many people asked whether Gram Panchayats would not simply fall into the hands of the dominant Lingayats and Vokkaligas? The answer to this last question is that most of the time dominant castes did get elected. But as the President of the Ford Foundation commented, 'the politicisation of once oppressed castes had reached a point where it was no longer easy to dominate these village bodies in all places at all times'. Further, 'initial evaluations of this system show that the elected councils have been more responsible to their constituents needs than the appointed administrators were under the previous system.

The district and sub-district councils have also played an important role in nurturing new political participants, especially women and members of disadvantaged castes... it seems clear that the concept of democratic decentralisation has been proven effective in Karnataka and deserves to be considered more widely'. *Should there be party politics at the Gram Panchayat level?* Whether or not political parties should be allowed to participate in Panchayati Raj elections is an issue over which even firm supporters of the immediate implementation of PRIs are divided. The Panchayat acts in different states do not say anything about the official participation of political parties.

The Sarvodaya notion of PRIs inspired by Gandhi and Jayaprakash Narayan believe that political parties have corrupted village societies. Jayaprakash Narayan believed that political parties hinder local self-government and initiative by making the poor and illiterate pawns in the party contest for power. The ultimate aim of the Sarvodaya movement was to encourage partyless democracy built on Gram Swaraj. In fact around 1961 Jayaprakash Narayan even advised the states to give rewards to villages which had elected Panchayats without any contest.

The other argument is that Panchayati Raj elections held on a party basis, will heighten the politicisation of people and give them choice.

In Karnataka the 1987 Panchayat elections were held along party lines. Individuals who had dominated the villages on the basis of their existing social and economic power were challenged by candidates put up by national or regional parties. Some studies have also found that often the harassment of women and scheduled caste members of the Gram Panchayat only became public because it was taken up as an issue by opposition parties.

Consensus candidates will only further perpetuate existing power structures in the village. Mathew has commented of Karnataka, Andhra Pradesh and West Bengal that 'it has now been clearly established that local bodies elections on party lines have made the villagers of these three states much more politically conscious, giving the elected Panchayat bodies a political standing of their own.

The local leadership has also become more politically responsible'. I asked several respondents about the effect of politics on PRIs and whether or not party politics is desirable at the village level. One respondent said 'elections without politics is wishful thinking. Politics are here anyway, better to make it formal and educate people about how it works'. Another respondent said 'people like to be connected to power bases at the centre through a political party.

It gives them the feeling that they have some power and are part of the system'. This perception is also evident in Andhra Pradesh; as Mathew's comments on local PRI elections suggest 'these elections were held on party lines. They evoked great popular interest throughout the state, generating heated debates on political personalities, ideologies as well as people's issues and were popularly referred to as the 'ini general elections' Another respondent commented: 'it is not only that party politics are working in Gram Panchayats. They are part of the local system. The recent elections to water user groups were fought along party political lines'.

COLLECTIVE ACTION AND EMPOWERMENT IN THE GUIDELINES

The Guidelines do not say anything explicit about collective action and empowerment. However the implicit notion of the dynamic between social capital and collective closely follows that which is provided by Putnam. That is that participatory development, through the construction of social capital, can lay the basis for effective government. The vision of community development in the Guidelines is similar to that being promoted in many natural resource programs today. That is the importance of incremental and sequential steps in the establishment of institutions.

Small Self-help Groups with common interests, linking into the larger community, with NGO or government partnerships providing a link between the group and the wider political and economic structure. The overwhelming evidence on CBOs is that they are dominated by the élite and that very few remain sustainable once the external input which motivated their formation is gone.

The combined evidence from respondents points to two enduring consequences that CBOs appear to have had on the empowerment of local people. On the one hand because CBOs are usually formed in the context of a project, local people have become dependent on external inputs. On the other hand, whilst this dependence may not appear positive, it has made people more aware of demands which can be placed. Further, CBOs have a lasting institutional impact in terms of building local awareness of possibilities, even if the organisation itself is not sustainable.

This section has argued that the current enthusiasm for building 'local capacity' and 'local institutions' often does not distinguish between different types of local processes. WIs and PRIs are mentioned in the Guidelines as interchangeable institutions for watershed management, when in fact their roles and functions are very different. Both systems have different inherent potential for stimulating local empowerment and collective action.

THE PROJECT INTERFACE BETWEEN PANCHAYATI RAJ

THE OPERATION OF THE SYSTEM FROM THE STATE TO THE DISTRICT

The operation of PRIs and WIs varies drastically from state to state. Karnataka has had a relatively long and successful experience of Panchayati Raj since its radical legislation in 1987. The Panchayati Raj system now operating is less radical, but Karnataka was still one of the first states to pass its conformity act to the 73rd Amendment and is still considered to be fairly progressive.

The Zilla Parishads are centrally involved in the implementation of the Guidelines. Andhra Pradesh has also had a relatively progressive experience of revitalising Panchayati Raj in the 1980s and holding elections. However although it has passed its conformity act, PRIs have not been given a very central role, and are not involved in the implementation of the Guidelines.

The two states also differ in their experience of implementing the

Guidelines. Karnataka has had a long history of watershed management projects, but has been slow in taking up projects under the Guidelines. Andhra Pradesh on the other hand is way ahead of all other states in the scale, scope and rate of watersheds adopted under the new Guidelines.

Karnataka

The administrative system for the implementation of the Guidelines in Karnataka is very decentralised. The state remits the programme funds to the district level, in the appropriate centre-state share, and is thereafter responsible principally for monitoring the programme. Monitoring is basically an auditing procedure of money spent and tasks achieved, with a heavy bias on physical targets.

The state sometimes intervenes if there are disputes between different institutions involved in the implementation of the programme, such as DWDAC and the PIA, but this is rare. The Secretary of the DRD and Panchayati Raj said it was a policy decision to decentralise such daily operation. Consequently there is minimal information available at the state level (either on paper or in person) on more qualitative aspects of how the Guidelines are being implemented. The main executive function at the state level is the convening of the SWIPRC once a year under the Chairmanship of the Deputy Commissioner. The various line departments and NGOs involved in watershed development are invited to attend this meeting to discuss operational and logistical issues, in particular those related to coordination between them.

The district level is therefore the principal executive and administrative centre for the implementation of the Guidelines. The structure of the district administration in Karnataka is unique because the District Rural Development Agency has been abolished and merged with the Zilla Panchayat. The President of the Zilla Panchayat is called the Adhyaksha. He is indirectly elected by the other elected members of the Zilla Panchayat and has the ultimate responsibility for planning development for the district.

The routine administration of development programs and projects, such as the Guidelines, is done by the Chief Executive Officer. S/he is an IAS officer and in charge of coordinating the various line departments. The CEO subordinates at the Taluk Panchayat level are called Executive Officers, and at the Gram Panchayat level they are called Secretaries.

The administrative and the political system are therefore closely intertwined in planning development at all three tiers in the district. The power relations between the CEO and the Adhyaksha are finely balanced. The CEO is able to object to any decisions taken by the Zilla Panchayat if he can show that they are against either the spirit or the letter of any law. Recent legislation in Karnataka has given the state a time limit to respond to such objections. If the matter is not decided then the Zilla Panchayat can proceed with its decision.

In the 1987 Panchayati Raj Act the Adhyaksha was responsible for the appointment and transfer of the IAS staff serving the Zilla Panchayat. The fact that transfers are no longer under the jurisdiction of the Zilla Panchayat was considered a big step backward by one MLA. It means in effect that IAS officials are not accountable to the people whom they are serving.

Many respondents stressed that the good relations between the CEO and the Adhyaksha are critical for the operation of the Zilla Panchayat. Both signatures are needed for the implementation of most decisions and for the release of funds. Because the daily functions of the two are different, it is possible for them to work in cooperation. However it is also easy for the relation to develop into a power struggle, driven for control over each other, and autonomy from each other.

The main district level responsibility is the appointment of the District Watershed Development Advisory Committee (DWDAC) and the PIA, and the financial accounting for the programme. Few operational decisions, and very little active monitoring or evaluation is done at the district level. For the operation of the Guidelines the procedure at the district level is fairly straightforward.

The CEO is in operational control and he constitutes a DWDAC of the selected PIAs, NGOs, and relevant line departments. Its objective is to provide a forum in which all of those involved in implementing the Guidelines can exchange experiences. The DWDAC is meeting, by all accounts, as per the Guidelines. It is well attended and afforded with considerable importance. The Adhyaksha of the Zilla Parishad reviews the decisions taken by the DWDAC during the sub-committee meeting on finance and planning which convenes once a month.

Most of the field level monitoring of the system and operational decisions are taken at the taluka level. The responsibility for this is devolved to the Executive Officer who sits within the taluka Panchayat. The Executive Officer is directly involved in checking sites and actively monitoring the progress to ensure that the work is being carried out.

These progress reports are then sent up the hierarchy, ultimately to the GOI. The elected members at the taluka level have more direct involvement in the operation of the Guidelines than the elected members in the Zilla Panchayat. Although this remains in essence an administrative task, rather than a strategic or political one, interviewees at the Zilla Parishad in Bellary said that the routine appraisal of programs by elected members contributes towards accountability.

Andhra Pradesh Andhra Pradesh has been one of the most progressive states in implementing the Guidelines and contributing to policies on watershed management. The Hanumantha Rao Committee incorporated many of its suggestions after examining watershed management in Ananatapur district in particular. Andhra Pradesh has implemented 750 out of the 850 watershed programs sanctioned and is widely recognised as being at the forefront of watershed development.

The reason for this appears to be the synergy that has been generated between a relatively politically conscious public, a relatively responsive political system, and a relatively accountable bureaucracy. Another reason is that Andhra Pradesh, especially under the Telugu Desam Party, has struggled to maintain some distance from the centre in the implementation of development projects. The State Level Coordination Committee decides on how and where to implement projects within months.

Andhra Pradesh can cheerfully bend and amend the Guidelines to local needs and priorities. Other states follow, or try to, every letter of the Guidelines. The result is a rapid turnover rate of projects, the attraction of more funds, creation of more capacity, and more projects. Panchayati Raj Institutions are not very active in Andhra Pradesh and have not been given the mandate to implement the Guidelines.

Andhra Pradesh had Panchayat elections in 1996 and issued its Conformity Act with the 73 rd Amendment, and has a party in power that is officially committed to Panchayati Raj. A reform of the Panchayati Raj system, and devolution of powers, was part of the election manifesto of the Telugu Desam Party in Andhra Pradesh.

When it came to power the Party took drastic measures to change the defunct Panchayati Raj system. In the 1986 Act it abolished 330 Panchayati Samities and established 1,104 Mandal Praja Parishads in their place. Each Mandal Praja Parishad is controlled by a group of villages with a population of 35,000 to 50,000. Reducing the size of the middle unit by one-third was intended to take the administration closer to the people.

However the reluctance of the 500 MLAs to devolve power has prevented the Amendments being implemented in the spirit with which they are intended. The PRIs are therefore not operating as institutions of self-government in Andhra Pradesh. PRIs are heavily dominated by the bureaucracy and, apart from education, have been assigned very few functional tasks.

The District Collector remains a member of the Zilla Praja Parishad committee and has the power to suspend or remove any member of the Panchayati Raj body including the chairperson of the Zilla Praja Parishad. The government retains the right to suspend PRIs and they have not been given sufficient work or the financial powers with which to implement them. Although PRIs have been constrained in their operation, there has been considerable political and administrative support for CBOs. The DRDA in Andhra Pradesh is widely recognised as being very efficient when compared to other states.

The Chief Minister has supported the establishment of CBOs for all sorts of activities, and has publicly expressed his views that political power as represented by PRIs has a corrupting influence. At the same time there is recognition that PRIs are ultimately important, and that there is a need to build up the capacity of local people to manage PRIs. Some work is being done with the Gram Panchayats through Janma Bhoomi and there is a move to involve Gram Panchayats in watershed management.

Subjectively, I would say that the government staff in Andhra Pradesh seemed to put more emphasis (than government staff in New Delhi or Karnataka) on building CBO capacity as part of a wider strategy for empowerment that would lead to development. A respondent in the Department for Rural Development commented 'CBOs are a kind of kindergarten for Panchayati Raj'. *Comments* In Andhra Pradesh the implementation of the Guidelines has been facilitated by good working relations between the administration and the political leadership.

The combination of a political leadership with a particular vision of development through CBOs, and a relatively accountable bureaucracy has provided the DRDAs and the WIs with strong central support. The political leadership has given the Department for Rural Development considerable freedom to select Project Directors who have the skills required for the work. The routine political pressures which influences many transfers, and often results in badly qualified and poorly motivated staff, have been eased. Although operational decisions are taken at the district level, the respondents I met at the state level had centralised information and a good grasp of the routine status of watershed projects. This centralised information, apart from the bare basics of money spent and projects sanctioned, was not available in Bangalore.

Decentralisation of watershed management to district level is more complete and the role of the state in operational decisions is more limited. The developmental role of the state and the links that have developed between the political leadership and the administration for the implementation of watershed programs is certainly a major reason for the rapid uptake of programs in Andhra Pradesh. This is probably part of the reason that there were fewer complaints in Andhra Pradesh than in Karnataka about logistical obstacles, such as delayed payments. This state level support is probably the biggest difference between the two states.

Apart from that, the fact that watershed programs in Karnataka are being implemented by the Zilla Panchayat does not appear to make a big difference. Essentially, it is still the administration (through the CEO and his subordinates) that is in charge of implementing the programs. They provide a delivery system, as they would if they were an independent DRDA, and are subject to many of the same constraints and incentives.

The fact that the CEO works in the ZPs does place some constraints on how the Guidelines are implemented, although it is hard to specify the effects of these. The system is still in transition, the legal relations between the CEO and the Adhyaksha are evolving, and the balance between them is heavily affected by personal negotiations over their role. The Andhra Pradesh approach is interesting because government support for CBOs has been part of a sustained programme, and because it is a programme that is supported by an articulated philosophy. Therefore, is Andhra Pradesh the right state to ask: does it work?

Are CBOs a sufficient basis with which to stimulate and sustain development? I will return to consider these questions in the following section on the community level. As far as the implementation of the Guidelines themselves are concerned, there is nothing that can help in the central question of the links between PRI and WIs, because there aren't any. The Guidelines are being implemented by the DRDA, and the institutional system - SWIPRC, DWDAC, PIA, WDT, WA and WC - has been more or less established. A review of this system can be found in Turton et al. (1998a).

THE INTERMEDIARIES: NGOS, PIAS AND PRIS

The lack of centralised information means it is impossible to get an overview of which institutions are PIAs in Karnataka. It is also difficult to assess the relative contribution of NGOs, the government, and PRIs, in implementing the Guidelines. The working relations between the government and NGOs in Karnataka have taken some time to develop. The problems (conflicts in working styles, time frames, approach, incentive structures, operational areas) are typical of those which have been recorded all over India.

The fact that the DRDA has been abolished, and that NGOs have to work with the Zilla Panchayats, has not made much difference to these conflicts.

The lack of centralised information, and the shortage of time for this study, meant that trying to understand the operation of the Guidelines was akin to guessing the shape of an elephant blindfolded. However it appears that the constitution of the DWDAC, the selection of PIAs, the establishment of WDTs, and the whole procedure for establishing the hierarchies from there downwards, is being done as per the Guidelines.

Copies of the Guidelines abound and at both the district and state level there was a comprehensive understanding of their contents. Almost all of the PIAs for the Drought Prone Area Programme (DPAP), Desert Development Programme (DDP) and Employment Assurance Scheme (EAS) projects are NGOs. In fact at the state level one interviewee in the government commented that the Guidelines are basically a programme for the employment of NGOs.

The PIAs for the Integrated Wasteland Development Programme (IWDP) tend to come from the government, which means one or two line departments under the control of the Chief Executive Officer. This is because the IWDP does not operate at the village level but starts with the whole district as its principal unit of operation. The working relationship between the government and NGOs depends on whose perception is solicited on this question. Several respondents commented that five years ago it would have been unthinkable for the government and NGOs to cooperate on projects as they are doing now.

Government employees are being seconded to NGOs and vice versa. The suspicion that characterised the relations between the two is disappearing as projects are designed which draw on the comparative advantages of both. There is a general acceptance, from both NGO and government respondents, that many NGOs are corrupt. Karnataka has a large number of NGOs who have been blacklisted. Of course no NGOs will include themselves in this category, and so from their perspective the government appears to go through lengthy selection procedures.

There is also a perception in the government that NGOs should stick to capacity building, whilst the government should be responsible for any physical work. This perception was expressed both at the state and district level. Several NGOs have applied to be PIAs, but have been offered only the 'drum-beating' preparatory part of the project. Having said this, other NGOs, such as Myrada, have been offered more work than they want to take up. In their opinion the procedures for monitoring and evaluation by the Zilla Panchayats, basically giant audit sheets, are too superficial.

They are reluctant to accept more work until they feel that the programs are being monitored properly. In Bellary the NGOs and the ZPs work closely together on watershed development, as well as all the other projects under the CEO. The ZPs has monthly meetings with NGOs to discuss future and on going projects. These complaints and differences of opinion over the respective roles of NGOs and the government are the same in Karnataka as they are in other states.

The fact that the Zilla Panchayat is in control does not appear to make much difference. However several respondents mentioned that NGOs are not satisfied about their working relations with the PRIs, because they ask too many questions. The Panchayats at the Gram, Taluk and Zilla level require NGOs to be accountable and have created a forum within which people can demand evidence. Several NGOs in Bellary, and the Deputy Secretary for Development, said that a 'three-way watch dog' system has developed between the government, PRIs and NGOs. One NGO member in Bellary said 'at the Taluk level elected members are asking lots of questions. Some rebel members are pushing for explanations for every decision'. Whilst this was generally considered positive, the NGOs also complained that the motivations for much of these questions were narrowly political.

The future working relations between PRIs and NGOs depends on the extent to which democratic decentralisation is actually pushed through. The Karnataka Watershed Development Project (KAWAD) in Bellary will be interesting to observe for this reason. At the moment, whilst this evidence of three-way accountability is encouraging, the relations between the ZPs and NGOs are more or less the same as if it were the DRDA.

Despite questions which may be asked at the Taluk or Gram level, the accountability of the NGOs is upwards, to the district, state and centre. There is quite a wide variation in the ways in which NGOs work with PRIs. In Bellary, NGOs attend monthly planning meetings and some elected members may attend these. However in Bellary the relations between the CEO and the Adhyaksha, apparently for personal reasons, are very good. In most districts NGOs coordinate and collaborate with the CEO and his staff.

The general perception of the other half of the Zilla Panchayat is they are elected fellows, they will not have any interest in watershed development. Most of the NGOs work with PRIs at the district level, because they have to, and coordinate with Gram Panchayats at the local level. However there are fewer NGOs who are implementing watershed projects and working with Gram Panchayats as an explicit policy decision to strengthen local democratic institutions. Most are working with Gram Panchayats either because it is practical to do so, or because they cannot avoid doing so. I came across no NGOs in Karnataka who were working directly with elected members at the taluka or Zilla level.

The general perception appears to be that progress at the Gram Panchayat level is possible, but at the taluka and Zilla levels the game is basically a political one which has no bearing on development. This seems a missed opportunity, as the development plans for the whole district are formulated at the district.

The experience of an NGO, PRADAN, in West Bengal is interesting to consider in this context. The PRIs in West Bengal have taken complete control over the implementation of the Guidelines. From its work in the Purulia district of West Bengal, PRADAN has recorded the difficulties it has faced in establishing its role in watershed projects. Some of these difficulties have been positive, in that they have been required to demonstrate the effectiveness of their methods.

The Sabapatti of one Panchayat in which PRADAN was working gave them one day to demonstrate the purpose of Participatory Rural Appraisal. However although PRADAN's work has been appreciated by the Panchayat, PRADAN has not been given a secure institutional niche in watershed development. 'As of now PRADAN does not have a legitimate space, recognised by the government'.

GRAM PANCHAYATS AND COMMUNITY BASED WATERSHED INSTITUTIONS

The Guidelines and Panchayati Raj, are based on different philosophies of how to generate collective action for development. This section will consider the evidence on the comparative advantage of Gram Panchayats and Watershed Committees in the implementation of watershed projects. There are two main questions that this section seeks to address:

- What are the comparative advantages of PRIs and WIs for implementing watershed programs?
- Can PRI and WI complement each other or are they mutually exclusive?

The following sections will consider the performance of Gram Panchayats and Watershed Institutions, differences in their roles and functions, and their effectiveness in addressing issues of equity. *Performance of Gram Panchayats and Watershed Institutions* Although the

Guidelines specify that Gram Panchayats are also entitled to take on the responsibility of implementing a cluster of watershed projects in the capacity of project implementation agencies, if they so desire, there are very few Gram Panchayats operating as PIAs or WCs. Watershed Committees and associations are formed from what will be the user group for the watershed.

The procedure for this follows a common pattern in Karnataka and Andhra Pradesh. If possible the WAs were established drawing on pre-existing SHGs or CBOs which had been working in the village. If there weren't any, SHGs are usually started to implement the work of the WA.

It is too early to assess the success of the WIs in implementing the works, ensuring equity in benefits, and sustainability. The only comment available at the state and district level in Karnataka on the performance of WIs is that the money is being spent and the structures are being built. The fieldwork done for this study was also insufficient to count as an evaluation of how well WIs are performing. The following comments are therefore drawn from the evidence gathered in interviews and papers.

The performance of WIs has to be judged against the two complementary aspects of watershed management projects specified in the Guidelines: 'improving the economic and social condition of the resource poor...through more equitable distribution of land and water resources...and greater access to income generating activities'.

To this can be added the implicit objective of empowerment through participation in development projects. The respondents had considerable doubts about the evidence for a sustainable improvement of the economic and social conditions of the resource poor. Most of the projects have contributed greatly to income generation, and the benefits of these activities do appear to be relatively just in their distribution of benefits to the poor. In many project areas, migration, an ongoing problem in the improvement of drought prone areas, has almost stopped.

Watershed management to most people means employment generation because most people do not have the time for soil and water conservation. If the project has good results then sometimes people get convinced of the value of soil and water conservation and do not only think about employment generation. The record of WIs in managing a sustainable improvement in the economic and social conditions of the resource poor through a distribution of land and water resources is more mixed. A recent study of projects that have been ongoing for several years could not identify any significant benefits.

The review of the Guidelines also found employment generation to be one of the major benefits and could not identify a clear improvement in the social and economic status of the poor. One of the major problems in establishing such benefits is that watershed management involves everyone in the watershed, which usually includes comparatively wealthy and large landowners.

Their preference will be for projects that will benefit their land that is often situated in the valley, and they will be able to influence the WCs to concentrate on such projects. In any case, many of the poorer farmers with marginal lands do not object. They spend most of the year either working as labour on the land in the valley or migrating, so they do not have a big objection to watershed management work which will benefit the land in the valley as long as they are paid. Another problem in watershed management is the difficulty of developing common lands.

The common lands, almost always seriously degraded and situated in the upper slopes, are particularly important for the livelihood of people with little private land. A healthy upper slope is also an important component of soil and water conservation further down. The development of this potentially productive land is impeded by several factors. First, most people will prefer working on private land, as the distribution of benefits from common land are always an uncertainty.

This uncertainty is exacerbated in most places by encroachments and complicated mixture of de jure and de facto property rights. Secondly, the wealthier members of the WCs are usually particularly opposed to working on common land, as they do not depend on these. Finally the opportunity cost of developing these is often too high for poor people, as the commons may have to be closed for afforestation.

To summarise, improving the social and economic conditions of the resource poor through a redistribution of land and water resources is difficult because firstly, it is a physically difficult task, and secondly, such moves are blocked by the élite when they feel that they could stand to lose either directly (diversion of water) or indirectly (benefit flows from a project going to others).

To understand why this is so, when PIAs are meant to explicitly target the poor, brings me back to my main question that is an understanding of the roles of PRIs and WIs and the links between them. Much of the physical work and capacity building for watershed management is done in SHGs, often composed of women or the poor. The attendance and performance in these groups is often appreciated as being impressive. However the final decisions on watershed management are taken in the WA and WC, in that the rich are also represented. The review of the Guidelines found that women in particular, are poorly represented on WCs.

Attaining a consensus here is much more difficult. As a contractor from Madanapalle in Chittoor district commented 'we build the tanks, bunds, whatever else is required with the participation of the whole watershed. If the rich people do not like the effect they simply knock it down again afterwards. Who will complain?' The celebrated successes of watershed management are usually those in which an NGO has had a sustained input over a period of years.

The Rural Development Trust (RDT) in Anantapur, for example, has allegedly achieved a sustainable process of involvement in many villages. Myrada has managed to transform (from what I saw on a video) a barren watershed into a green watershed that is so productive that the decrease in migration has become a sustainable fact. The secret of the Myrada success, apart from being an effective NGO, is the commitment to generating participatory development and the sustained input of time and resources. They started by creating Self-help Groups for credit and then took up other activities such as watershed management.

The SHGs were strong enough to influence the WCs to take up work which benefited them directly, and Myrada contributed with technology which was able to harness benefits for all without too many costs being imposed on any particular group. These indicators of success prompt the question of the sustainability of such an approach. Aloysius Fernandez of Myrada argues that for watershed management to be equitable and sustainable the participatory process will take five years, with no short-cuts.

Myrada has in fact been working in its successful watersheds for much longer. Several reviews have noted poor participation to be a principal reason for project failure. They have recommended that projects have a start-up phase of capacity building before watershed management work starts, and possibly a 'sustainability phase' as a follow-up.

The move towards asking communities to make a contribution towards the work done in watershed management is one positive step towards ensuring local commitment. *What of the Gram Panchayats and their performance in implementing watershed programs?* As mentioned above there are very few Gram Panchayats who are acting as a Watershed Committee or as PIAs. I came across Gram Panchayats who have undertaken watershed development schemes, but these are not on programs that work through the Guidelines. Out of these there were a few success cases. For example, a respondent from the Karnataka watershed development Society related a case where ten Gram Panchayats had contributed Rs.40 million towards watershed management.

Development Alternatives, an NGO, is working with a Gram Panchayat in Tumkur district, that has expressed its willingness to do the entire watershed management work on a loan of 5 lakhs, and to repay it all within an agreed fixed term. I asked members of the Dasudi Gram Panchayat in Tumkur district whether watershed management was an issue that came up in their development plans.

The Sarpanch said 'We receive one grant a year from the government, and watershed management is not a priority with that money. But let the government give us a loan and we will do watershed management'. He said that one of the major problems for Gram Panchayats is that they have little autonomy to act and that most decisions have to be approved by the district or state. They had been waiting a year for permission to carry out drainage sanitation work. 'After one year nobody is interested in doing the work any longer'. These examples reveal that Gram Panchayats are able to support watershed development.

However such extreme cases are the exception rather than the rule, and most NGOs are reluctant to work with Gram Panchayats. Four reasons are usually cited for this reluctance. The first is that the Gram Panchayat is too big a unit and its boundaries rarely coincide exactly with those of a watershed. Secondly, Gram Panchayats are too heterogeneous and are dominated by élites.

Thirdly, Gram Panchayats pursue short-sighted political goals and these usually revolve around more immediate tangible projects such as house and road construction. Fourthly, and underpinning all of the above, is the perception that the Gram Panchayat and the Sarpanch do not represent the Gram Sabha.

A respondent in Anantapur said 'the Gram Panchayat is a one-person institution. He is indirectly elected by the other Panches and given a mandate by them to do as he pleases. So the Gram Panchayat is useless unless the Gram Sabha gets a mandate for development'. This opinion is entirely representative of everyone I spoke to (the difference in opinion as I will discuss below, lies in the solution to this problem). Despite these problems, in both Karnataka and Andhra Pradesh the link between PRIs and WIs is strongest at the village level.

The WAs and WCs usually include a few (1–3) members from the Gram Panchayat. The Sarpanch is an honorary member of these associations, but cannot be President. The logic behind this system is to include them in decision-making and to benefit from their power, but not to allow them to take over.

A respondent from Anantapur commented 'that is how we mesh the two systems together. The Gram Panchayat members are not in control, but they are also not out of control'. It is often the case that the members of the Gram Panchayat automatically become members of the Watershed Association because their land lies within the project watershed. In such cases, the chances are that they will be larger landholders so the benefits that they can bring to the WAs, in terms of official support, have to be weighed against the possible disadvantages of their priorities in watershed management for smaller landholders.

The reason for including Gram Panchayats, or forming links with them, varies from project to project. The reason given by many (over half the respondents involved in implementing projects) for forming links with Gram Panchayats was functional. They were contacted because they could not be easily avoided, for example: Gram Panchayats are contacted for permission when common land is taken up for afforestation or soil conservation work.

Most PIAs I spoke to said that this was a routine procedure and that permission was usually given, although some PIAs have had problems with the Gram Panchayat withholding the land. Gram Panchayats are also on occasion contacted for logistical support, such as, information on administrative procedures or opening accounts. At other times they are contacted in order not to antagonise powerful local interests. Of more interest for this study are the instances when Gram Panchayats are contacted for the explicit purpose of building sustainable links between PRIs and WIs. I will consider these in the following section.

EQUITY IN PRIS AND WIS

What is the vision of the village community, caste and gender in PRI and the Guidelines? How are they able to contribute towards empowerment of the poor? PRIs and WIs have different approaches to empowering the poor. Both take the poor to mean, broadly, scheduled caste, scheduled tribes and women. The Panchayati Raj Act specifies affirmative caste and gender action of a 33% reservation for both legislative and executive offices (Article 243–c and d). The objective is to reserve seats for them at every level of the hierarchy thereby ensuring that their needs are represented at every level and that policy can be made at the national level, which is sensitive to these needs.

The Watershed Guidelines provide for such affirmative action only at the level of the User Group and provide a normative guideline for the committees. The WIs approach is to build the capacity of these groups by implementing projects that are relevant to their needs and motivating them to participate in these and thereby gain the confidence to plan others.

The emphasis is on fostering the empowerment of the poor within groups especially created for that purpose: for example all women SHGs. The Guidelines have three limitations in promoting equity in watershed management across the country. The first is that they do not provide an institutional mechanism with which to scale-up this empowerment or a process whereby the poor can enter mainstream institutions. The second is that watershed management projects are increasingly being targeted at communities that have shown a willingness to cooperate. Whilst this is a sensible move for a project, it also avoids conflict resolution, and therefore abandons many people who might stand to gain from watershed projects.

Finally, most watershed projects are still land-based and have not seriously tackled the problem of how to build up the livelihoods of the landless or land poor. Common property areas have also received less attention than private land, which, assuming that the link usually made between poverty and dependence on common land is correct, biases projects towards those with more private land.

The PRIs are revolutionary in the institutional support that is given to the poor, and the opportunity for power that it presents to them. But how is it working in practice? Many of the issues involved, such as those surrounding collective action and the formation of social capital, have already been discussed in this report. As always the theme returns to the Gram Sabha. This is the key to the PRI system and as long as it remains weak, the whole system will be a farce. There has been little research done on if and how the Gram Sabha is changing. However the evidence collected in the context of examining watershed development suggests that some rapid changes are taking place. One respondent described these changes 'the élite know that if they are not careful they will kill the golden goose'.

He was speaking about broad changes in the village community, including both PRIs and WIs. In Hirschman's (1970) formulation discussed: opportunities for voice. PRIs and WIs are contributing towards a process whereby disadvantaged people can negotiate instead of either remaining loyal or taking radical exit measures. A respondent from Orissa commented that in both Andhra Pradesh and Karnataka there is a system for demanding benefits from the state, whether this is through the Zilla Parishad or the DRDA. The benefits may not get evenly distributed, but the system is there and people are learning to demand some accountability from it.

An academic who has worked in Andhra Pradesh, Karnataka and Orissa compared the two. 'Karnataka and Andhra Pradesh are ten years ahead of Orissa in administrative reforms. The people there are able to demand benefits and the state has to provide them with these. It is not working perfectly, but in Orissa it is not working at all, everything is still completely supply driven. It takes time for the system to work as it will swing along pendulums.

In Karnataka the system after 1987 was working very well with the Adhyaksha having power. Now they have less power, but the pendulum will swing back again'. In Karnataka there was a general perception that the experience of Panchayati Raj has had a positive effect on local understanding of politics and their rights.

The same opinion was given in Andhra Pradesh, although the vehicle for social change was more often put down to the naxalites, government moves on transparency, the media, and general economic mobilisation. An NGO respondent in Bellary commented 'the weaker sections are at a learning stage now, but this is the chance for the government to help them. In the next five years they will start coming up. They are learning how to mobilise for funds. The next elections will be very tough'. The experience of such mobilisation is of course extremely diverse. I visited two adjacent Gram Panchayats with very different perceptions of what PRIs could achieve. In Hoyisalakatte the Gram Panchayat had not done anything substantial for years. Both the political and the administrative system were perceived as a single hopeless corrupt system.

Those people who are strong enough to change the system just wait until it is their turn to be corrupt. In Dasudi the verdict on the operation of the Gram Panchayat was less bleak. Several of the women from the SHGs were considering applying to the Gram Panchayat for loans to support vermicompost businesses started under watershed development programs. One woman, a poor and abandoned Muslim woman, had despite her marginal status succeeded in pressing the Gram Panchayat for support. In both Gram Panchayats the potential for at least momentary inspiration about Panchayati Raj was evident when they remembered 'Nir' Sahib.

'The time when Nir Sahib was in control was great for Panchayati Raj. We were all prepared to make sacrifices for our village'. The effectiveness of reservations as a means to ensure empowerment of the poor has been a much- debated issue. Most of the Gram Panchayat members in the village belonged to the élite, although both had a woman Sarpanch, in one case a scheduled caste woman.

The sentiment expressed at the meeting (the latter Sarpanch was not present) was that reservations were bad because they allow inefficient people to take control. This is of course the classic case against reservations, and it is not surprising that it is expressed by Brahmins and Kshatriyas about a scheduled caste woman. From the evidence on the field trip, and in reports on reservations, the barriers against members of the scheduled caste/scheduled tribe appear to be much stronger than the barriers against women.

The violence of this barrier was terribly evident recently in the Tumkur district when a scheduled caste man was made to eat human excrement to quell his political ambitions. Gram Panchayat work at the moment is almost wholly focussed on what benefits can be harnessed from the Taluk and district. This does make it very difficult for a poor person with very local knowledge to be effective.

A politically astute person who can make the right contacts and bring development schemes to the village, may in fact be better in the short-term. The poor in this Gram Panchayat were not aware of the system and they relied more on each other for assistance. For example, in one tribal village called Dabbagunte within Hoyisalakatte Gram Panchayat, a respondent commented 'there is a strong community feeling in the village and we help each other. There is also a pride in not asking others for assistance'.

A member of the Gram Panchayat commented 'such communities are perfect for politicians because they can get their votes in one block and they don't need to give the people anything in return'. As for the scheduled caste woman President: 'she doesn't understand anything'. I did not meet her, but some people from her household said that she is not happy in the position and that she does not have enough time to work now. I met the other woman President and asked her what motivated her to become involved in the Gram Panchayat: 'women have always been active in this village, and we have never received any problem when we give our opinions.

Now it has been formalised, that is all. Before we would not have had the idea to have a formal position'. I tried to interview the female Adhyaksha of a Zilla Parishad, but was told by her office that she was at home. When I phoned her home, her husband answered and said that his wife does not go to the office unless there is something to sign, in which case they send for her by car. There are abundant examples of the failures of reservations. Does this mean that they are a bad strategy? What about success cases?

First, the positive influence that women have on shifting development priorities to sustainable 'soft' development issues (such as soil and water conservation, afforestation, sanitation) are now well-known. This was evident in the Gram Panchayats I visited where it was women who were pushing the Gram Panchayat to be more active in this respect. There is plenty of evidence that this effect is there even when women's representation is only nominal and the husbands are effectively signing the papers. Secondly, empowerment for women is often a very rapid process and Community Based Organisations have been an important part of this empowerment process.

Thirdly, caste divisions in elections are fading and being replaced by coalitions that span caste barriers. Building more formal links between the two systems is one way in which the 'capacity' built through WIs and other CBOs can be scaled-up. There are many instances of these linkages between CBOs and PRIs emerging. In Andhra Pradesh 800 women from SHGs contested the Panchayat elections of which 400 were elected.

A respondent said that such a turnout would have been impossible if CBOs and SHGs had not enabled women to recognise their own capacity. At a recent Telugu Desam rally a women Sarpanch from Warangal district publicly asked the Chief Minister why there were no women in his circle of advisors. In the Myrada area 54 women who were part of SHGs have been elected to PRIs.

The pattern is the same in the Khribhco areas of Gujarat, although the numbers are smaller. The process from SHG to PRI often happens without any external encouragement. Government representatives in Andhra Pradesh were especially conscious of the success of this process and the rapidity with which it is happening. The habitation committees of the Janma Bhoomi project also consist of members of CBOs and Gram Panchayats. It is one of the most interesting findings of the linkages between PRIs and WIs and is a process whose potential needs to be explored further.

Many organisations are already involved in assisting the poor to enter the Panchayati Raj system and to voice their demands. However most of these organisations are advocacy based, there are very few field-based NGOs amongst them. For example, ActionAid has recently released a CD-Rom on women in Panchayati Raj. United Nations Children's Fund (UNICEF) supported a telephone and television conference via satellite in which elected women from all over Karnataka could share their problems and ask questions to a panel in Bangalore.

PRIs and WIs have different objectives, different visions of the community, and a different perspective on the process of collective action. It is not necessary for these differences to be a point of conflict as long as policy is formulated which respects the designated roles of these two systems. The conflict is likely to arise if these differences are brushed under the carpet.

PRIs are units of self- government which have been given a constitutional mandate to plan development, raise revenue, open bank accounts, spend money, and, importantly, to place democratic pressure on a system which is linked to political power at the top of the system. The vision of the community is not of an autonomous and atomistic local body, but of one which is affected by external influences and decisions taken in the wider polity, and therefore has the right to be part of these decisions through the democratic process.

Watershed management is one of the functions under the domain of the Gram Panchayat in the XIth Schedule. Gram Panchayats are in fact meant to develop a plan for the management of all the natural resources within their boundaries. However the jurisdiction of the Gram Panchayat is too big to manage micro-watersheds, and with all the other issues listed in the XIth Schedule, the Gram Panchayat would also not have sufficient staff or salaries for such management.

There is therefore clearly a need for a Community Based Organisation to manage watersheds in a locally sensitive manner with a local negotiation of costs and benefits. This is the role of Watershed Institutions. There is no conflict if WIs aim to empower local people, educate them about their rights, or provide them with economic tools for a better life. Nor is there a conflict in the roles of PRIs and WIs if the latter chose to set up federations at regional or national levels.

The problems may arise when the roles of the WIs take on those of the Gram Panchayat and when WIs start planning the broader development process. The problem arises first of all because the Guidelines were not designed for this purpose: the WIs have no legal status, no representation at higher levels, and no statutory stability.

Again there is not technically a problem with this until the WIs are supported as an alternative system to PRIs. As Arora notes; 'sporadic efforts have been made to federate SHGs/CBOs at the sub-district and district levels. These groups are independent organisations and can therefore federate at whatever level they want but if they are supported by development projects which are either donor funded or state funded, then one must examine whether such federations are likely to conflict with higher level PRIs.

The fall out of such a conflict is likely to be faced by the state and so whether or not the state should finance a higher level federation has to be carefully assessed'. There seems to be confusion about the roles of PRIs and WIs, when in fact they can be compatible. One of the main reasons given by respondents for not involving Gram Panchayats is that they are too big for watershed management.

This is confusing a technical matter with a principle. The appropriate size for WIs is an on-going issue of debate and is usually negotiated anew with each project: should it be based on a micro or macro watershed? Should all the households in a watershed be included or only those belonging to a particular village.

Similarly, the appropriate size of the Gram Panchayat is much debated: are they not too big to be representative of the needs of the village? Are they not too small to incorporate economies of scale in development plans? These logistical problems are not a reason for treating these institutions as if they are interchangeable. Whatever the size of a Gram Panchayat, WIs can flourish within them, as long as they are ultimately accountable to the Panchayati Raj system. In Karnataka the legislation for the establishment of societies within Gram Panchayats was established by an Amendment in 1997.

Indian Food and Fertiliser Company, the agricultural cooperative, is already working with parts of different Gram Panchayats on this basis. The success of PRIs depends heavily on the effective functioning of the Gram Sabha. But again, there is not necessarily a conflict as the evidence included in the Gram Panchayats on the success of CBOs getting marginal members suggests. Further, it is a mistake to think that WIs can forever be free from the influence of politics and power. 'You know what will happen when the project staff withdraw?

The élite people will be waiting to crush the group. Their protection will be gone and they will be helpless'. This one step forward, one step back, was the very negative response of one project worker. It may be more a case of two steps forward, one step back, but there is certainly plenty of evidence of project gains being rolled back. Further, it is not possible for WIs, especially with the amount of money coming in, to be insulated from power and politics.

Watershed Association members are often coopted by political parties that are attracted by their funds. The recent elections to the WCs in the coastal districts of Andhra Pradesh, for example, were fought on party political lines. It is better that these politics are made explicit and transparent. The roles and functions of the WIs and PRIs also differ in their ability to raise revenue and tax.

The PRIs have been given, in theory, the powers to raise revenue and collect taxes. In practice the ability of Gram Panchayats is usually restricted, and most state finance commissions have not yet specified the subjects which will be handed over to the Gram Panchayats. Since the Amendments were passed there has been some pressure on the state to devolve down to the Gram Panchayats all the subjects that are in 'their natural domain' for taxation. At the heart of this statement is the principle that an argument has to be made for decentralisation, rather than vice versa. One of the first areas that would fall in their domain is the collection of dues for all of the natural resources that lie within their boundaries. These are presently being collected by the Revenue Department and the Forest Department in particular.

The Gram Panchayat would become an important nodal point for planning, as well as having the resources to invest in development. At the district level it has been argued that the PRIs have short-time horizons for planning and are likely to favour projects which have immediate visible benefits. This may be true, but it has to be compared with an administrative system which also has rapidly changing fads, and whose staff are transferred every year.

The District Collector in Anantapur, for example, has expressed his interest in water harvesting, and is not interested in the soil and water conservation work that has been the foundation of much of the work done by NGOs in the district. This lack of continuity, and the lack of local knowledge, is a frequently voiced criticism of the state administration. A respondent in the Department for Rural Development in Bangalore pointed out that the roles of the CEO and the Adhyaksha complement each other's expertise. 'The Adhyaksha is a local fellow who knows about local conditions and has to please the local people. He may not be having experience of how to implement programs. The CEO has experience of running development projects, but he may not be having local knowledge. It is a good combination. There should not be any conflict'.

Chapter 10

Accountability and Sustainability

One of the defining differences between PRI and WIs is the principle of accountability on which they operate. PRIs are accountable to the electorate and the sanction is being voted out. WIs are accountable to the government, who can stop funds and disband committees if procedures are not followed as specified. PRIs seek to endogenise the mainstream political process to ensure its accountability. The Guidelines seek to exogenise the political process as this is seen to subvert accountability. The question of accountability is therefore at the heart of the operation of the political and administrative system. It is also at the heart of the question of what links exist, and should exist between the two systems.

Accountability is a central component of the good governance agenda and a frequent political promise in India, as elsewhere. The good governance agenda identifies several important components of accountability: accounting; decentralisation and NGO involvement; and micro-level accountability of public agencies to the consumers of public services.

As Moore notes these are listed without asking or answering the questions; 'accountability to whom? Through what mechanisms? And to what degree?'. It is not at all clear, for example, why there should be a connection between NGOs and accountability. Accountability 'is a highly abstract concept, sometimes interpreted in formalistic and legalistic terms, and sometimes used in a more concrete way to refer to the social, economic, political, etc. mechanisms through which some agents become responsive to other agents'.

Moore suggests that the obscurity, in which accountability is enshrouded, prevents an examination of the mechanisms through which it could operate. This appears to be the case for both PRI and WIs. The former is enshrouded in so many Amendments, provisions and ordinances that, as one respondent put it, 'the law itself has become a moving target. How can we possibly implement it?' The latter has an elaborate system of monitoring and evaluation that disguises the fact that the accountability is to the top: governments and donors. A deconstruction of the concept of accountability is imperative for an understanding of the political and administrative systems for development projects. I will start backwards, by listing my conclusions, and then provide the evidence from the fieldwork to support these.

- Decentralisation without a devolution of executive, financial, and judicial powers is frequently tantamount to, at best 'passing the buck', and at worst, 'decentralising corruption'. This is true for both the political and administrative systems.
- Panchayati Raj Institutions will not work unless they are made accountable to the Gram Sabha. The Gram Sabha cannot work unless it is given power.
- Watershed Institutions are accountable to the Zilla Parishad given the legislative remit of the?Panchayati Raj Act.
- Traditional bureaucratic mechanisms of accountability should not be downplayed in favour of a?vague accountability to 'the people'.
- Authority and accountability must be linked.

A deconstruction of accountability is useful because it provides a means through which to test the claims of Panchayati Raj and the Guidelines. In Karnataka there was a widely felt and voiced opinion that 'the system' was not accountable to local people. There was no distinction made between the political and the administrative system in this respect.

The problem with the Panchayati Raj system is that the Gram Sabha has not been sufficiently politicised and has not been vested with any judicial, executive or policing powers. 'We cannot do anything if the Sarpanch is not distributing any benefits. What should we do, you tell us?' To my response, vote him out, came the reply, 'the next one will be the same'. In the other Gram Panchayat the people were less pessimistic, but the general feeling of apathy with the system appears pervasive.

The reason for this lack of accountability is that the political system, the Zilla Parishad, has basically become a channel for the delivery of development projects to the village. They are not accountable for the delivery of these projects to the village but to the donors or higher hierarchies in government. In this sense the Zilla Parishad operates just like a DRDA and has the same lack of accountability. The monitoring and evaluation procedures are basically enormous audits of money spent and structures built. The physical monitoring is done by IAS staff at the taluka level and by the WIs but there is no check on the process or quality of the work.

Myrada has declined taking on more work because of this lack of monitoring 'they require us to fill in enormous forms but there is no check on the quality.' It is important to note that the reason for this sort of accounting is not because it is a Zilla Parishad.

The procedures are the same with a DRDA, and in any case, the Guidelines are implemented and monitored by the IAS within the Zilla Parishad. There is also very little accountability and uniformity to the way in which the WIs are established. Government members of staff are often suspicious of NGOs and many NGOs have been blacklisted for bad work and corruption. They commented that NGOs are against PRIs because they are beginning to ask questions about NGO performance.

NGOs concede that there are many corrupt cases, but point to the corruption and lack of accountability, within the government. One NGO/PIA commented 'we are refusing to pay and that is why we will not get any more work. If you really want to see how the Guidelines are working then don't spend only two days. Stay here for a month and go to each and every project site and look for the work that has been done.

You will find exactly nothing.' He had written a letter to the CEO and the Adhyaksha accusing them of corruption and has consequently received no further work. There are checks and balances between the PRIs and the state administration. This is true especially between the Adhyaksha and the CEO, and several respondents said that this is working well because both become accountable to each other, and both have the power to put obstacles in each others way. The perception of their relative power at the state level in Bangalore was also mixed. For the purpose of the Guidelines the Zilla Parishad is generally considered in terms of the CEO and his officers.

There were some fears at the state and district levels of passing funds down to the Watershed Committees. 'They are not accountable to us. It is very difficult to control them. But the taluka level is maintaining a tight control.'

One respondent talked about a 'three way watch dog' system between PRIs, NGOs and the state administration. They also mentioned competition between NGO and government PIAs as having a positive effect because both want to prove that they can perform better. The NGO respondent said 'for years we have been saying that we can do the work better and more cheaply than the government. Now that we are PIAs we want to prove it.' However whilst this system may improve efficiency it will not necessarily improve accountability.

Respondents within the PRI all commented that the system has taken some backward steps from the Panchayati Raj Act that was introduced by the Janata Dal in 1987. The PRIs themselves are now less accountable to the lower levels, and the state administration is less accountable to PRIs.

One of the biggest problems is that PRIs no longer have the right to request the transfer of inefficient IAS officers. A former Adhyaksha commented 'this is a real step backwards, it is terrible. It means that the state administration is totally unaccountable, and that if there is an inefficient officer there is nothing that can be done about it.' Most respondents I spoke to did not think that the PRI system was inherently less likely to be accountable than the state administration. 'The state has had its chance for 50 years, why not let the PRIs have their chance. They may be as corrupt but they can't be more corrupt.'

A former Secretary at the Ministry of Agriculture commented 'the irregularity with PRIs is far less than with the government. The Zilla Parishads are already handling large amounts of money, so why not trust them with implementing the Guidelines?' It is difficult to judge the accountability of the PRI's system as it has never been left in place long enough to be tested.

One of the reasons given for aborting the Karnataka Panchayat Act of 1987 was that PRIs are corrupt. However there is much dissent on this conclusion. No one disputes the existence of corruption, but most argue that it is less than what occurs in the government. Further, it is argued that it is the people who should ultimately judge this. 'If a case is built up by using charges of corruption in order to scrap the system or diminish the importance of people's representatives, this is tantamount to throwing the baby out with the bath water.

After all, in a democracy, isn't it the people who should punish their corrupt and wicked representatives as and when the next elections take place? Instead these very elections were postponed indefinitely. What kind of logic is this?'. *Andhra Pradesh* The question of accountability in Andhra Pradesh is more broadly interesting because of the high profile campaign for transparency that has been pursued by the Chief Minister.

Whether this transparency also ensures accountability is not a question that this report can answer. However two tentative conclusions can be drawn from a consideration of the structure for development projects in Andhra Pradesh. The first is that a positive synergy between the political and the administrative system is important. As one respondent put it 'the political mindset does influence the bureaucracy.'

The second is that a good bureaucracy is important, and traditional systems of bureaucratic accountability should not be bypassed in favour of an undefined accountability to the people. The IAS cadre in Andhra Pradesh has a good reputation and there is considerable prestige attached to working in the Andhra Pradesh cadre.

There is an attempt to pick the right person for the work in Andhra Pradesh, as opposed to just allotting positions through routine transfers, and all the usual influences on these. For example, the Secretary of Rural Development has made a recent request to select fifteen Collectors as Project Directors for the Watershed Management Projects, and has received no political opposition. This is very unusual. The Andhra Pradesh IAS cadre has had a good reputation for being accountable since the 1980s but it has come to the foreground with the campaign for transparency pursued by the Chief Minister.

For whatever reason (personal or the indirect circumstances under which he came to power, his background in social mobilisation, his pursuit of the female vote,) the Chief Minister has placed great emphasis on 'bringing the government to the door of the people'. Some examples are his Chief Minister's question time, (he can be asked anything by anybody) and his file clearance week, (when all pending issues are cleared).

He has also promised to computerise governance down to the Mandal level so people have easy access to information. At the village level a system of monitoring and evaluation has been introduced, with a prominently displayed social audit board detailing information on projects. Whilst some of these measures have been criticised as partly gimmick, there is no doubt that it has a tremendous influence in contributing towards people's empowerment. The Eshwaran Committee recommended a quiet revolution through a change in the roles and responsibilities of the bureaucracy, and in Andhra Pradesh, it seems that this is moving in a generally positive direction.

The link between authority and responsibility is more tangible in Andhra Pradesh than in many other states. Conventional systems of ensuring bureaucratic responsibility are important, and even a fully functioning Panchayati Raj system cannot replace this, also it can put pressure on the administration to instate this. Jain (1985) notes about state sponsored development programs that, 'monitoring and evaluation (which were rigorously applied during the 'community' development and 'Panchayati Raj' phase), are now conspicuous by their absence.'

The public and prompt availability of reports of the Programme

Evaluation Organisation in the past was distinctive of the Community Development Programme. Such a public monitoring system is critical to the success of a programme and Andhra Pradesh has made considerable moves to ensure that it is put into place. However, whilst monitoring and evaluation are important, they are finally only technical tools. The critical issue is one of accountability.

Although the state administration in Andhra Pradesh is more transparent than in other states, there is still little local accountability. At a workshop on Perspective Planning for watershed development, attended by the government and NGOs, the question of accountability came up several times. Several NGO representatives said that they had no idea of what funds arrive at the district level and how they are spent.

They felt it was important to have some knowledge about the scale of the programme. Others commented that monitoring and evaluation is still a numbers game, a giant audit, with little reflection on the success of watershed projects in targeting the poor. It is acknowledged that in Andhra Pradesh, as in most others states, the Watershed Committees are dominated by the resource rich who receive a disproportionate share of the benefits.

A respondent from RDT commented that 'approximately 60% of the Watershed Committees are not democratic institutions', but are dominated by the powerful. In particular, contractors have a tremendous influence on the use of the resources within the WC. It is important to recognise that there is an informal and covert structure of accountability that underpins the distribution of resources from development projects.

The government and civil society agents are often constrained in their operational flexibility by the demands placed on them by client groups. In conclusion, the record of accountability to the 'beneficiaries' of development projects is poor in both Andhra Pradesh and Karnataka, (although better than in most states) most likely because they are at the moment only 'beneficiaries'. Accountability can only be achieved when members of the Gram Sabha and beneficiaries of development projects are given more rights, and are made aware of these.

At a recent conference on Perspective Planning for Watershed Development in Hyderabad at the National Institute for Rural Development (NIRD) June 1–3, 1998, this question of local accountability was not considered. There were proposals to set up an apex body to oversee watershed development programs which would be accountable to its stakeholders: the government, NGOs and donors.

But no mention was made of local people, and in any case to include them in the list would be avoiding the issue. At the central level some initial steps to look into ways of ensuring local accountability are being taken by the DRD in New Delhi. One possibility is through the institutionalisation of Jan Sunwai, the system for public hearings. In the meantime the campaigns for transparency in Andhra Pradesh and the right to information movement in Rajasthan are a step in the right direction.

The latter movement has been a campaign sustained by the Mazdoor Kisan Shakti Sangathan. 'The Panchayats will break this centralised information system when 30 lakh elected members ask for information on a variety of matters that affect the people's lives. It is hoped that the work being done by Mazdoor Kisan Shakti Sangathan in Rajasthan will spread to other states and that the Panchayats will be the most effective vehicles for it' Mathew and Nayak (1996). If the movement was part of the reason that the GOI published the Right to Information Act in April 1998 then it has already had some effect, although there are so many provisions to this act that it is virtually unworkable.

SUSTAINABILITY AND 'SCALING-UP'

One of the most debated issues in watershed management projects is the question of how to ensure that the project is sustainable and that the benefits are 'scaled-up' to other areas. The question of the sustainability of PRIs, in theory at least, because most of them are not functioning, is a moot point; they are constitutional and they will exist as long as the democratic system is perceived to be a good one. WIs are project specific and task specific organisations.

Their sustainability will depend, political factors aside, on the support that they receive, and the extent to which local people feel they are necessary. The sustainability of the institution is not important. What is important is the sustainability of the process which they have contributed towards, that is an improved livelihood basis and the construction of 'social capital' through enabling the poor to act collectively and independently to find their own solutions.

The overwhelming evidence on natural resource management projects, such as watershed management, is that the benefits are not sustainable in the long-term. Plantations disappear, committees are disbanded or abandoned, the livelihood base of the poor remains only marginally improved, if at all. Indeed the record of the sustainability of most government projects is extremely bad. Perhaps in some cases they create some sustainable social capital by raising awareness amongst the poor. However the evidence suggests that most of the social capital which projects create is held by the rich, and that in fact most projects have contributed to a further entrenchment of their powers.

Part of the reason given for NGO involvement in development projects is to ensure the sustainability of a project by threading it into the fabric of the local social and economic structure. Some NGOs have ensured the sustainability of some projects by creating participatory development and negotiating conflicts before leaving the village. This approach takes time and needs resources, hence the question of how to 'scale-up' this process to cover wider areas. Further, these success cases are drops in the ocean, and even then they are not problem free.

A respondent from RDT, an NGO with a very good reputation commented that 'there are a few villages in which the committees have the capacity to become partners of the government for development. But 60% of the institutions are not democratically elected.' Another study of natural resource management projects found that after NGO support ranging from 7–12 years, 'the social organisations or community groups involved do not appear to have reached the stage yet where external support – whether operational or 'institutional' – is no longer required'.

The question of sustainability and scaling-up featured prominently in a recent workshop on Perspective Planning for Watershed Development held in Hyderabad. A consensus was reached that watershed management is a necessary but not a sufficient basis for improved local livelihoods.

Watershed management needs to be linked to other activities, especially for the landless, to ensure its sustainability. The concept of 'Watershed Plus' has emerged to describe this process. The precise meaning of the term is still being negotiated, as reflected in the exploration of this term by two different discussion groups at the Hyderabad workshop. One group included broader development objectives, such as; equity, gender equity and poverty alleviation as components of Watershed Plus. The other focussed more on the means; credit, employment, organic farming, drinking water.

Between them they covered (apart from health) a substantial part of the rural development portfolio of any country. With the identification of such broad objectives, the next logical step in the discussion was the question of how to scale-up the capacity to implement such a programme.

One immediate conclusion was that the existing institutional capacity was insufficient. One NGO commented 'we need additional capacity for capacity building, we need capacity building for mobilisation and we need capacity building for execution. RDT does not have even two free days to contribute to this process 'we are already extremely busy'.

The various suggestions which arose on how to solve this bottleneck, in scaling-up all involved building more institutions, recruiting more people, and building apex bodies at various levels to monitor these building capacity to identify capacity. The discussion on the institutional capacity to oversee this process focussed on resources within the government and NGOs. One idea is that these apex bodies would identify a representative in the village.

This approach to sustainable watershed management raises several questions about the links that exist, and should exist, between PRIs and WIs. What should the role of PRIs be once watershed management becomes a broad rural development programme rather than a narrowly technical one? Should PRIs not be involved in coordinating these issues? There are several reasons to suppose that this would ensure the sustainability of watershed management. First, they are responsible for development planning for the whole Gram Panchayat, and would be the natural apex body for linking watershed development into other objectives. Second, they are democratically elected, so they can reflect local needs and priorities as they change.

A third critical issue is that it will be more cost effective for PRIs to be the apex linking body as they were already established for this function. The infrastructure is already there, they have the power to raise revenue, and labour will be cheaper. Many small NGOs mentioned that they were not able to implement the programme under the current financial allocations and an increased allocation of funds was a starting assumption for all of the programs planned at the workshop in Hyderabad.

The project implementation and fund disbursement rate for watershed programs is rapid in Andhra Pradesh. However as a respondent from the Department of Rural Development in Delhi said 'the DRDA in Andhra Pradesh is working very well at the moment because it is a bureaucratic body flush with funds.' Whether or not this is sustainable remains an open question. The Panchayati Raj system is by its nature sustainable. Whether they will be effective remains to be seen. However many NGO respondents have already started working with them because, in the words of a respondent from AFPRO in Hyderabad: 'what alternative do we have?'

Comparisons between states in India could be very useful because it means variations within a common framework. This report does not constitute the basis for a thorough comparison. However, there are several implications that can be drawn from the links between the political and administrative systems in watershed management on the basis of this comparison.

The comparison between the two states is implicit in the report, so I will just briefly summarise them in this section. First, the general experience of popular mobilisation has made a difference in both Karnataka and Andhra Pradesh. In Andhra Pradesh the Telugu Desam Party and its demands for more regional autonomy, the naxalite movement, as well as the periodic revival of PRIs, has been part of this popular mobilisation.

In Karnataka, the Janata Dal and the experience of Panchayati Raj after 1987 has made an indisputable difference to local awareness. A MLA respondent in Bangalore commented 'there are 80.000 PRI functionaries, and they are beginning to demand things.' The experience and character of popular mobilisation is subject to political swings: in Karnataka the provisions of the 1987 Panchayati Raj bill have been diluted, and in Andhra Pradesh the next government may be less enthusiastic about support for CBOs.

However, the gains that have been made cannot easily be withdrawn, because an awareness of rights is true social capital, and is not attached to any particular institution. Sunil Khilnani has commented on democracy in India that; 'democracy as a governmental form will no doubt suffer the vicissitudes to which all human institutions are prey.

But as an idea, as a seductive and puzzling promise to bring history under the command of the will of a community of equals – a promise that, given the inevitable gap between intentions and consequences, can at best only hope for partial fulfilment – it has irreversibly entered the political imagination'. This perspective appears a little hopeful when standing in the tribal village of Dabbagunte, that routinely delivers their block of votes and ask for nothing in return.

However, the combined opinion of the respondents in the interviews suggests that it will not be possible to roll back all of the promises made to the electorate in Karnataka and Andhra Pradesh. The second point that can be made by comparing watershed management in Andhra Pradesh and Karnataka is that in both states watershed management is still an externally funded activity.

Comparing the accountability of the Zilla Panchayat and the DRDA reveals a common constraint in both the administrative and political systems. Essentially both systems, whether DRDA or the Zilla Panchayat, are delivery systems. There are slightly different influences operating on how and where they deliver, but they remain delivery systems, dependent on resources from above.

Local development planning, whether through the Gram Panchayat, the WCs, the DRDAs or the ZPs, is still limited, although some of the Self-help Groups have begun to take autonomous initiatives. There is a need to make the system more accountable and to deepen and expand the scope of the democracy in both Andhra Pradesh and Karnataka.

This report has identified three questions that are critical to an exploration of the links between Panchayati Raj and the Guidelines. These questions are:

- What kind of decentralisation is necessary for watershed management?
- Does the involvement of PRIs in the implementation of the Guidelines require institutional change or involve institutional conflict?
- How can one support local collective action and how do Panchayati Raj and the Guidelines compare in this respect?

Decentralisation Panchayati Raj and the Guidelines are both systems that aim to decentralise control over development to local communities. The former is a constitutional part of Indian democracy and the latter is an executive order of a ministry. Apart from the fact that they both aim to decentralise, they differ in their legislative status, objectives and approach.

All of the respondents interviewed stated that the 73 rd Amendment, which made Panchayati Raj a constitutional part of Indian democracy, was long overdue and is progressive legislation. Further, that ultimately PRIs are a fundamental and sustainable part of local empowerment. The respondents differed on whether they thought that PRIs were ready to take on this responsibility, and on whether Gram Panchayats are an appropriate unit for watershed management.

The usual objection to Gram Panchayats being made responsible for watershed management, is that they are too big, and don't coincide with watersheds. This report concludes that these debates about 'the perfect fit' between the physical watersheds, and the social units that should manage them, miss a fundamental point.

This point is that the two systems were established for different purposes. The PRIs are not implementing committees or agencies for projects, but units of local self-government. It is more than likely that a Gram Panchayat will not coincide with a watershed, but it does have a mandate to make a plan for natural resource management within its boundaries.

There is no contradiction in establishing Watershed Committees within a Gram Panchayat. However, when Watershed Committees receive GOI and donor funding to expand their roles, and take on many of the functions assigned to PRIs, there is a need to be careful of not undermining the respective roles of the two systems. Much of the recent enthusiasm over 'Watershed Plus' has led watershed management into territory which is in the domain of PRIs.

Whether or not this creates a conflict is highly dependent on local and state politics. However in all cases it is a matter of missing the 'spirit' of the legislation underlying both Panchayati Raj and the Guidelines.

Further it misses out the comparative advantages of both systems. In the case of the Guidelines, this is Committees especially established for watershed management with a support network of line departments, NGO expertise, and specialised supervisory teams. In the case of Panchayati Raj, this is a sustainable local institution that has the statutory power to plan natural resource management, open bank accounts etc., and represent local needs at higher levels.

INSTITUTIONAL CHANGE AND CONFLICT

One of the underlying questions in this report has been whether there is a conflict between Panchayati Raj Institutions and the Guidelines, and whether this will need institutional change to resolve. There are several areas where the functions outlined in the Guidelines are similar to those outlined in the Panchayati Raj legislation. In themselves, these do not constitute a significant issue or problem. However they are part of the larger process of the development of Panchayati Raj and administrative decentralisation.

This process of institutional change has started in many states, and involves changes that will be politically negotiated. The negotiation is likely to occur between sectors (government, NGO); within the government (centre, state and district); and between the PRIs, the state administration and NGOs. The Guidelines refer to institutions as ideal types.

It is important therefore to appreciate that they are based on already existing institutional alliances and conflicts, that institutional change is usually politically resisted, and that any 'negotiation' will take place between groups within uneven resources. There is also unexplored room for cooperative work between PRIs, NGOs and Watershed Institutions at various levels.

Many respondents felt that the development of Panchayati Raj Institutions, particularly Gram Panchayats, will lead to decreasing NGO involvement. This may be true in the very long-term, but it will take some time before PRIs are operating effectively, and many NGOs have accumulated skills in 'capacity building', which are needed by PRI members.

The description of the role that NGOs and voluntary associations should play in the seventh plan, is in fact one of facilitation, supplementing government work, and capacity building. Planning development, which is part of the domain of PRIs, is not included, although many of the larger and apex level NGOs have in the last ten years been informally drifting into the policy planning process. Finally there is room for cooperative work between Gram Panchayats and Watershed Committees at the grassroots.

Power structures within a village are fluid, and as all grassroots workers are aware, it is not possible to change these by setting up new committees. SHGs can sometimes manage to insulate themselves from local politics, especially if they are composed mainly of women and receive little funding. However as soon as committees receive funds they attract local power groups. The evidence of this process is overwhelming, and is well demonstrated by the recent Watershed Committees elections in Andhra Pradesh which were fought along party political lines.

It is therefore difficult to draw a line between Gram Panchayats and Watershed Committees, and maintain that the former is for politics and the latter for watershed management. However, the roles of the two institutions are formally complementary. Many NGOs have made significant progress by supporting weaker groups to establish their needs in the sustainable and formal mainstream of village politics. *Local collective action* Panchayati Raj and the Guidelines have very different approaches to supporting local collective action.

In Panchayati Raj local collective action is formalised through the democratic process and unprivileged sections of the community are drawn into the mainstream of political life. In the Guidelines, collective action is issue based, and unprivileged sections of the community are supported to make a united stand based around specific needs. The question of that is more effective, is at the heart of the debate over whether to support Gram Panchayats or Watershed Committees and how to construct the links between them.

The implementing agencies I talked to all had a slightly different approach towards supporting local collective action for watershed management. Most were opportunistic, building links between Gram Panchayats and Watershed Committees based on the local reality. Some NGOs were more strategic, and had taken an active decision either to work mainly through SHGs, or to have the strengthening of the Gram Panchayat as an ultimate goal.

In all cases the comparative advantage of SHGs, Community based Organisations and Gram Panchayats was a central discussion point. This report concludes that both approaches are important, and that ultimately comparing their potential for watershed management is futile.

As discussed above, their roles in watershed management are different. Further, many of the discussions about the comparative advantage of Gram Panchayats versus Community Based Organisations, become embroiled in a comparison of the track record of these institutions. This exercise has limited utility. It is now fairly clear that CBOs of all kinds need sustained external support, and that most are not particularly representative of the needs of the poor. It is also evident that Gram Panchayats are dominated by the élite, and that so far they have been used to entrench élite interests.

The ultimate objective is local empowerment and development, and both types of institutions have an important role to play in facilitating this. However, this involves a political decision to follow the 'spirit' of the legislation in both Panchayati Raj and the Guidelines.

There is an urgent need for action research on the grassroots of Panchayati Raj: the Gram Sabha and the Gram Panchayat. Many of the academic respondents commented that there has been so little comprehensive research on the Gram Panchayats that the whole debate over how they function is being steered by anecdotal information. There is therefore a serious knowledge gap on the process of local politicisation and local understanding of the Panchayati Raj system. The 73rd Amendment has inspired a renewed emphasis on action research which combines an understanding of the constraints faced by the Gram Sabha with support for possible solutions: this approach requires support.

There is a growing network of institutional support for the Panchayati Raj system. Part of this network includes organisations that have taken a policy decision to support Panchayati Raj. Part of the network includes individuals who support Panchayati Raj within organisations (the government, donor agencies, and NGOs) that have not taken an official decision to actively support Panchayati Raj.

Of the former, organisations that have taken a policy decision to support Panchayati Raj, advocacy based NGOs and academic institutions are most active and vocal. In fact the link between these two types of organisations is often very close, with individuals simultaneously involved in both activities.

The Institute of Social Studies (ISS) in New Delhi produces annual updates of the Status of Panchayati Raj in different states after the 73 rd Amendment. It also publishes a newsletter called Panchayati Raj Update on the routine operation of Panchayati Raj; undertakes detailed research studies and conferences/debates on Panchayati Raj. Many other academic institutions (such as the Institute of Social and Economic Change) in Bangalore have departments and individuals whose passive support for Panchayati Raj over several decades has been rejuvenated by the 73rd Amendment.

Of the issue-based NGOs, many of the NGOs who work on gender, sanitation, health and education have started to work actively with Gram Panchayats. Perhaps part of the reason is that in many states, these are amongst the first activities handed over to the Zilla Parishads.

Further, the statutory involvement of women and scheduled castes and tribes in the system has motivated many of these NGOs to support the role that they will play in the system. ActionAid has recently brought out an excellent CD-Rom on the role of women in the Panchayati Raj system. UNICEF in an innovative attempt to provide support to recently elected women members of PRIs sponsored a television and telephone satellite conference in Karnataka.

Women from all over Karnataka were able to phone a panel in Bangalore to voice their concerns, ask questions and share their experiences. UNICEF has also produced a film called Sanshodhan (The Amendment) together with the National Film Development Corporation, about a woman who is elected to her local Gram Panchayat. Such awareness-raising activities as well as training courses and capacity building exercises are now increasingly also sponsored by the government.

For example, to celebrate Panchayati Raj Divas in Orissa in 1993, Biju Patnaik, the Chief Minister, invited the assembly of 16,000 elected representatives, almost half of who were women, of PRIs to Bhubaneshwar for a two day deliberation in the Kalinga Stadium. NGOs involved in natural resource management activities, whether this is watershed, wasteland, water or forest based, have been relatively slow in forging working alliances with Gram Panchayats. Many NGOs work with Gram Panchayats simply because they cannot avoid doing so, but have no active programme or ideology of support.

ActionAid has taken a policy decision to support Panchayati Raj through its network of NGOs in all activities. In Andhra Pradesh Oxfam, the Centre for World Solidarity, and Action for Food Production (AFPRO) are all actively exploring ways of working through Gram Panchayats for resource management in the future.

The Netherlands Organisation for International Development Cooperation (NOVIB) has supported a research study in Andhra Pradesh, Karnataka and Tamil Nadu to examine the links between PRIs and CBOs for natural resource management in particular. The Ford Foundation has also provided ongoing support on research studies and innovative experiments with PRIs.

The planned Karnataka Watershed Development Project will experiment by implementing the entire project in Bellary district through PRIs. Finally, NIRD in Hyderabad, at the initiative of the MRAE, has produced a suggested scheme for the devolution of powers and functions to Gram Panchayats, and has dwelt at length on neglected areas like land reforms, protection and maintenance of village commons and water management.

POLICY ISSUES IN MANAGING IRRIGATION AND DRAINAGE

F. Kennedy, the former president of the USA once said: "Anyone who solves the problem of water deserves not one Nobel Prize but two — one for science and the other for peace". This is a simple expression of the complexity and difficulty of managing water resources. Agriculture accounts for two-thirds of the world's use of fresh water. Some 40% of the world's food crops are produced by irrigated agriculture. The performance of irrigation and drainage is critical to the food supply, farmers' incomes and the environment.

This Bulletin aims to highlight major policy issues in the management of agriculturally used water resources, mainly irrigation and drainage. Before the discussion of individual policy issues, the Bulletin will link the relationship that exists between the characteristics of irrigation water; goals and problem areas of the management; and policy issues in managing irrigation and drainage.

CHARACTERISTICS OF WATER FOR IRRIGATION AND DRAINAGE

Water for irrigation and drainage has multiple characteristics. It is an economic good and also a social good. An economic good is a common material which can be traded as a factor of agricultural production, such as fertilizer. A social good is something that may have a benefit (or cost) to society beyond that which accrues to the individuals consuming it, such as low-cost housing. Irrigation water is another example, as it has social implications that involve a large number of mostly poor farmers.

Irrigation water can also be a private good and sometimes a public good. A private good, such as an apple, can be individually and exclusively consumed. A public good, such as national defence, has no excludability (i.e. it is difficult to exclude others from consuming it) and zero subtractability (i.e. many people may simultaneously consume it without decreasing each consumer's benefits). In the case of a large irrigation and drainage system, it would be very costly to exclude others from using it.

Furthermore, because of limited sources of water and limited sites for building structures such as dams and canals, irrigation-drainage is often a natural monopoly. Also, irrigation and drainage systems, particularly modern ones, have economies of scale. They are large in size and complicated in structure. This characteristic also limits market function. In addition, the effects of irrigation and drainage activities are highly interdependent with other uses of water and land resources, and have considerable and lasting impact on the environment.

GOALS OF WATER MANAGEMENT

The ultimate goals for managing water resources in general, and irrigation-drainage water in particular, are efficiency, equity, and sustainability. When the goal of efficiency is achieved, each drop of water has been properly allocated and used. This implies that the marginal returns of water for each user (rich or poor) for various purposes (rice or corn etc.) should be the same, and equal to the marginal cost (and price) of water. In such cases, the economic benefits of society from water can be maximized.

Of course this is an economic ideal which could only be achieved if both the water supply system and market conditions are perfect. In a practical sense, efficient water use means a proper allocation without waste of water. The goal of equity, on the other hand, refers to a fair distribution to users. For irrigation and drainage, some users may have more advantages than others.

Such a difference can be found between canal head users and tail enders, and also between influential farmers and poor farmers. In many cases, equity is more important than efficiency in irrigated agriculture. This is because irrigation efficiency is a difficult concept. Poor irrigation efficiency in the upper stream of the river basin may be beneficial to those down stream. However, unfair distribution to users in a system tends to be undesirable both economically and socially.

The goal of sustainability expects that current users maintain the quantity and quality of water resources for the use of future generations. It is extremely important that current water users keep a productive and healthy environment for generations to come. Poor water management in irrigation and drainage may have many negative effects on agricultural productivity, health and the environment. These effects include sedimentation, toxic discharges, water-logging, salinity, and even desertification.

PROBLEMS IN WATER MANAGEMENT

Because of the above-mentioned characteristics, it is very difficult to manage irrigation and drainage so as to achieve efficiency, equity, and sustainability. From past experience in many countries, there are three principal problems in management. These are market failure, government failure and environmental deterioration.

Market failure refers to the inability of market mechanisms or the invisible hand of capitalism to allocate water properly to different users in various sectors, or manage it efficiently without government intervention. Water sale and pricing are useful, but their function is limited. Water moves through a complicated hydrological cycle of rainfall, absorption, runoff, and evapotranspiration. This makes various uses of surface water and groundwater highly interdependent.

External factors such as pollution, waterlogging, and over-use of groundwater are often major causes of market failure. High transaction costs are another cause.

Measuring and delivering water, and monitoring the quantity and quality, to serve the market's purposes can be very expensive. The third major cause of market failure is the lack of secure and effective property rights over water for each user. This is particularly the case in irrigation and drainage in developing countries. Stealing of water by users is common. Moreover, on the supply side, the characteristic of a natural monopoly of irrigation and drainage is a fourth major cause of market failure.

Government failures refer to weaknesses of government authorities in managing water resources in general, and irrigation and drainage in particular. Given water's special characteristics, there are good reasons for governments to intervene, and even to directly manage various water uses by means of irrigation departments and water supply authorities. In many cases, however, governments may not have done this job well. Fragmented management is often the main cause of government failures.

Many governments create separate departments or agencies to manage different types of water use. In some governments, irrigation, the municipal water supply and water used to supply electric power, or as a means of transportation are each handled by different departments or agencies. Each is responsible for its own operations, and independent of the others.

This leads to uncoordinated and fragmented decision-making; excessive and unproductive investment; and undesirable environment effects. Problems are likely to be more severe when a water source involves various levels of government, such as central, provincial and local governments. Overextended government agencies are another major cause of government failure. There is often a heavy dependency on a centralized agency to plan, develop, operate, and maintain irrigation systems.

The agency is often too large, bureaucratic and inefficient. It hires too many members of staff, wasting precious funds on their salaries. Hence the cost of water supplies, particularly the operation and maintenance costs, are often too high, while the service provided is generally poor.

As a result, the users, who are mostly small-scale farmers, are often unwilling to pay. Furthermore, there is little user participation. This creates a vicious cycle, in which high costs plus poor service lead to users' unwillingness to pay, causing inadequate funds for operating and maintaining the system, leading to a further decline in the level of service.

ENVIRONMENTAL DETERIORATION

Many developing countries pay too little attention to water quality and pollution control. The discharge of untreated industrial wastes, the runoff of agricultural chemicals, and poor use of land for agriculture and forestry, all cause widespread degradation of land and water resources. In India, for example, a country that depends on irrigated agriculture, more than 4 million hectares of once-productive land have been abandoned because of water-logging and salinization. At present, about 10 to 15% of the world's irrigated land is affected by waterlogging. This is a common and difficult problem in agricultural water management.

MAJOR POLICY ISSUES

Policy issues here refer to various non-technical issues in managing irrigation and drainage. There is a long list of such issues. The following are among the most important.

- The need for a comprehensive multi-sectoral framework for managing water resources as a whole.

- The need for demand-driven management of irrigation for agriculture.
- The need for participatory irrigation management.
- The need to control water quality for irrigated agriculture.
- The need for poverty alleviation in irrigation management.

COMPREHENSIVE CROSS-SECTOR FRAMEWORK FOR MANAGING WATER RESOURCES

Current trends in the growth of population, urbanization, industrialization, and income have generated increasing demand for more water quantity and better water quality. As water is a limited resource, conflicts between various water users from different sectors and locations are increasing, and becoming more complicated.

It is almost impossible to optimize water management for the agricultural sector or any other sector on its own, without considering the needs of other sectors. The UN conference on environment and development in 1992 in Rio de Janeiro emphasized the need for a holistic approach to fresh water, and the integration of different sectors in water management, within the framework of national policy.

The World Bank states:

"Water resources should be managed in the context of a national water strategy that is based on an assessment of the country's water resources. The strategy would spell out priorities on water services; establish policies on water rights; water pricing and cost recovery, public investment, and the role of private sector in water development; and institute measures for environment protection and restoration..... In essence, this compre-hensive approach breaks down very complex problems in a river-basin into more manageable elements to achieve coherent cross-sectoral water manage-ment."

The World Bank approach views water as a single resource with many uses. The Asian Development Bank also emphasizes an integrated and cross-sectoral approach at a national level to avoid the consequences of uncoordinated and *ad* hoc development of water resources.

The cross-sectoral framework is a general map in managing a nation's water resources, in which a broad range of national policy issues will be involved. These include self-sufficiency in food, property rights and ownership, urbanization and regional development.

DEMAND MANAGEMENT FOR IRRIGATION AND DRAINAGE

To improve the management of existing irrigation and drainage systems, there is not much to do on the supply side except rehabilitation and modernization of these systems. These hardware investments can be very costly, and their impact is relatively short-term unless measures are taken to improve the demand side. Demand management is user-oriented. It involves mainly software activities, including regulation, education, pricing, incentives, user participation and re-allocation of water rights.

These activities influence water users directly or indirectly to adopt efficient water-use practices. Today, demand management has become the main way of managing water resources, particularly for irrigation and drainage, which involve a large number of small-scale farmers. It is difficult to improve the management of irrigation and drainage without cooperation and participation by these farmers.

Many further policy issues regarding demand management can be raised. These include the pricing of water; providing incentives for farmers to save water in the dry season and to drain water in the wet season; educating farmers in better management of water; and re-allocating water among sectors to adjust the long-term demand for water from various users.

Which measures should be taken depends on the country or region. The basic principle of demand management is to focus on the user. How to maximize users' participation in the full range of water management activities is the key in improving irrigation and drainage performance.

PARTICIPATORY IRRIGATION MANAGEMENT

Since 1994, the World Bank Institute (WBI) has carried out a programme of Participatory Irrigation Management (PIM). It has been very successful. PIM refers to the involvement of irrigation users in all aspects and at all levels of irrigation management. This includes planning, design, construction, maintenance, and financing for the main irrigation system and all subsystems. 'Involvement' is flexible, ranging from information sharing, consultation, and joint assessment to shared decision-making, collaboration, and empowerment. The rationale of PIM is mainly derived from market and government failures in managing irrigation and drainage, and the need for demand management.

The World Bank employs a new word, "userism", to express the essence of PIM. It is indeed management of the users, by the users and for the users. It is believed that PIM, or user management, can save financial and social costs compared to government management. Also, irrigation users have a stronger incentive to manage water more productively, and can respond more quickly to management problems in the system. In recent years, there have been many successful examples of PIM.

THE MEXICO EXPERIENCE

In the 1980s, the Mexico government suffered a serious financial crisis, resulting in a reduced budget for the maintenance of national irrigation systems. This created a vicious cycle in the irrigation sector — rapidly deteriorating systems, leading to farmers' unwillingness to pay for water, causing a shortage of funds which made systems deteriorate further. In 1990, an irrigation district of about 80,000 ha was transferred to the users. While the government manages head works and the main canal system, the users' irrigation association manages the remaining secondary and tertiary canal system.

By 1995, more than two-thirds of the nation's 3.2 million ha network had been transferred to 316 irrigation associations. Each association has a legal contract with the district water board. Management of the board remains in the government's hands for a specific share of the total water supply.

The contract, covering twenty years, specifies the proportion of irrigation fees that will go to the association and to the district water board. The so-called Mexico Model has saved the country from its crisis of irrigation management. In general, farmers were very pleased with the performance of the users' associations, and were willing to pay even higher water fees (without government subsidy).

THE CASE OF TURKEY

Following Mexico's lead, Turkey adopted a similar policy of irrigation transfer in 1993. More than half the national systems administered by the government in Turkey were transferred to local users' associations within the first three years. The results were encouraging. Farmers were paying roughly the same amount of water fees. However, a much smaller part of the fees were used for personnel. Irrigation services significantly improved. It was not possible to have irrigation during weekends and nights before the transfer, but this became possible when users managed the system.

THE STORY IN ALBANIA

In 1997, Albania suffered from serious political and social instability, which almost reached the level of a civil war. Most public infrastructure and facilities, such as powerhouses and pumping stations, were severely damaged. However, those irrigation systems which were managed by users were spared. Men, women and even children, the brave group of users, defended their irrigation facilities by force from rebel damage. Albanians called this "The power of PIM".

THE CASE OF ANDHRA PRADESH, INDIA

Following the example of Mexico and Turkey, the Legislative Assembly of the State of Andhra Pradesh approved an Act called 'Farmers' Management of Irrigation Systems' in 1997. According to this Act, the management of about 5 million ha was to be brought under the control of millions of farmers, mainly smallholders. Over 10,000 Water Users' Associations (WUAs) and 174 Distributory Committees were created.

They were given the responsibility of planning and implementing the maintenance and improvement of irrigation systems and water distribution.

The role of the irrigation department changed from a "doer" to a "facilitator". During its first year of operations, the area where the Act was put into practice reached 200,000 ha. Agricultural productivity in this area increased (INPIM 1999).

Participatory Irrigation Management (PIM) is not new and has been practiced in various forms in different countries for years. Japan's Land Improvement District and Taiwan's Irrigation Associations have existed for many decades.

Countries such as China, Vietnam, Pakistan, Egypt, Morocco, Nepal and Jordan are in the process of finding their own way to implement PIM reforms. The International Network on Participatory Irrigation Management (INPIM), an international non-government organization, was established in Washington D.C., with funding from the World Bank and some bilateral donors, to promote PIM reforms world-wide (INPIM 1997).

IMPACT OF PIM REFORMS

The impact of transferring irrigation management from government agencies to users' associations may be viewed from three perspectives: that of the farmers, that of the government, and that of the irrigation department.

The government is likely to be a winner, as subsidies for irrigation can be reduced. Farmers are also probable winners, as they can enjoy a sense of ownership with improved services. The irrigation department, however, may be a loser. Its budget, staff, and authority are likely to be significantly cut, and its role changed. IPIM is desirable and is likely to be the general direction most countries take in irrigation management. However, there are many policy issues in the planning, implementation and evaluation of PIM reforms, and different countries are likely to decide these differently.

WATER QUALITY MANAGEMENT AND WATERSHED PROTECTION

Water quality management is as important as the management of water quantity, and is even more difficult. This is because water quality management involves many external factors that can easily be neglected. Irrigated agriculture, in many instances, may have a negative impact on water quality and the environment. These negative effects include soil erosion, fertilizers, pesticides, herbicides, salinity, waterlogging, and flood runoff. Fertilizers add excess nutrients in the forms of nitrogen and phosphorous to streams and lakes. Pesticides and herbicides can contain toxic chemicals which may be lethal to wildlife. Water-logging and salinity will reduce land productivity, and also contribute sediment nutrients and chemicals to runoff.

Runoff from animal husbandry can add greatly to non-point pollution. Because there are many water users, most irrigation systems are large. It is difficult for governments to control water quality effectively. Both regulatory and market approaches are needed. In the regulatory approach, the government sets regulations (quality standards) that dischargers must meet.

This is basically a command-and-control format. The advantage of this approach is its simplicity and fairness, rather like setting a speed limit on the roads that applies to all vehicles. However, there are disadvantages. First, it requires expensive research and a long time to set quality standards. Second, high administrative costs are involved in enforcing the standards. Third, standards are often too rigid for individuals to meet, and sometimes discourage free enterprise. These disadvantages leave room for a market-oriented approach.

The market-oriented approach offers the promise of economic efficiency and user innovation. It may also be administratively simpler than the regulatory approach. One way to implement a market-oriented approach is to use the concept of the total maximum daily load (TMDL), to evaluate the total waste that a stream can assimilate while still maintaining its standard of quality. Once the TMDL is determined, then point and non-point sources can be allocated. This should make it possible to develop innovations such as pollution trading. This kind of self-regulation can help increase economic efficiency.

In addition, a watershed management strategy should protect the ecology of the watershed and control water quality. Many policy issues, however, should also be addressed. These include the land tenure system, the agricultural extension network, and incentive policies for credit, taxes and subsidies.

Secure land tenure will encourage farmers to invest in land and water for the long term. An effective agricultural extension network can teach farmers new technologies and practices, while incentives can induce farmers to do more to protect the watershed and reduce agricultural pollution.

POVERTY ALLEVIATION AND IRRIGATION MANAGEMENT

Poverty alleviation is a major goal of national development in most countries. What irrigation management can do to help poverty alleviation is an important policy issue. Among others, the following approaches may be considered. Treat the poor equally as irrigation clients. In many cases, resource-poor, small-scale farmers are disadvantaged water users.

They may be located at the tail end part of the system, or have little connection with water management officials. As a result, they often suffer from frequent water shortages, which make them even poorer.

To break this vicious cycle of poverty, the government and water management authorities must adopt a service-oriented management approach, and treat the poor as equal clients of the system. This should be the minimum that irrigation management does to alleviate poverty and achieve equity in water management.

Let the poor be favorably represented. Generally the poor are a silent group with little influence in irrigation management issues. Recently, there is an encouraging trend in Mexico of letting the poor be represented on the Board of Directors of the newly established water users' associations.

The Directors of the Board are elected separately from two groups of users—small farmers or tenants, and farmers with large land holdings. Directors from each group alternate as chairperson and other key positions on the Board. This helps protect the interests of the poor. The same approach can be used to solve conflicts between different social groups in water management.

Reduce the cost of water for the poor. There are various ways of reducing the cost of water paid by the poor. These include subsidies from the government, lower water fees for the poor, providing more jobs for the poor, or allocating favorable water rights to small users in dry seasons.

I have described five related policy issues in managing irrigation and drainage. They are:

- The need for a comprehensive, multi-sectoral framework in managing a nation's water resources.
- Demand management for irrigated agriculture.
- Participatory management for irrigated agriculture.
- The need to control the quality of water used for agriculture.
- The need to alleviate poverty in irrigation districts.

Turning these policy issues into action plans needs more research and training.

Policy research can formulate appropriate policies for specific issues in a specific place, while training can raise public awareness and upgrade the capacity to implement these policies.

I sincerely hope someone will be able to solve the problems of water, and receive two Nobel Prizes — one for science and the other for peace — in the foreseeable future.

NITRATE EXPORTED IN DRAINAGE WATERS OF TWO SPRINKLER-IRRIGATED WATERSHEDS

The increasing concern of society about the negative environmental effects of intensive agriculture threatens the expansion of irrigation in many areas of the world, including the semiarid south of Europe. However, the high solar irradiation and extended frost-free periods make these areas capable of high yields of field crops without deleterious environmental effects, provided proper management of irrigation and fertilization are used. Nitrogen is the nutrient that requires better management because it can be lost from the soil–crop system through runoff, leaching, denitrification, and volatilization. Since nitrate leaching is frequently the most important loss process in irrigated agriculture, adequate management of irrigation water is required for efficient use of nitrogen.

The amount of nitrate leached from a field is highly variable, being influenced by the irrigation system, soil characteristics, and climatic conditions. Since nitrate leaching imposes a cost on both the farmer and the environment, it is essential to quantify these losses and establish best management practices aimed at its reduction. Surface irrigation is the main irrigation system worldwide.

Water and nitrate losses below a crop's root zone are almost unavoidable in the conventional management of surface irrigation due to low efficiency and nonuniformity of application. This is one of the reasons why crops are overfertilized by farmers. Thus, nitrate losses greater than 100 kg N ha^{-1} yr^{-1} have been measured in semiarid irrigated areas in Spain and USA.

On the other hand, properly designed and managed sprinkler

irrigation systems allow for uniform and efficient application of irrigation water, which minimizes water and nitrate losses through deep percolation. In addition, the split and timely application of fertilizer N through the sprinkler systems makes unnecessary high application rates of N at planting and reduces the risk of nitrate leaching during the early growing stages of crops.

The aim of this work was to quantify the concentration and mass of nitrate exported in the drainage waters of two sprinkler-irrigated watersheds and to analyse their relationship to irrigation and nitrogen management in these semiarid areas.

DESCRIPTION OF THE STUDY AREAS

The study was conducted in the irrigated areas of Watershed A (from April 1997 to March 1999) and Watershed B (from October 1997 to September 1998), located in Hydrological Sector II of the Monegros II irrigated area (Ebro River Basin, Aragón, Spain). Both watersheds drain into the Ebro River through the Valcuerna Gully. The proper name of Watershed A is "D-IX", while that for Watershed B is "D-XI." The terms Watershed A and Watershed B are used in this manuscript for simplicity and clarity.

The hydrological limits of Watershed A are clearly defined and cover an area of 558 ha, which is drained through the 4.4-km-long Watershed A collector. The northwest hydrological limits of Watershed B are open and not well defined. We selected a study area delimited in the northwest by the concrete-lined Monegros Canal, except for two irrigated fields that were located on the west bank of the canal. The 1007-ha study area was drained by the 5.9-km-long Watershed B collector. The rest of Watershed B is dryland dedicated to winter cereals and fallow.

Three types of soils can be distinguished based on topographical position in the watersheds. The soils in the lowest areas are Typic Torrifluvents, generally deeper than 2 m, with alternating fine and coarse horizons, but mainly with clay texture. The soils of the hillsides (Typic Haplogypsids and Typic Torriorthens) are relatively shallow (<1.0 m), while the soils of the platforms (Typic Torriorthens) are very shallow (<0.5 m) with a clay-loam texture in both cases. The watersheds are underlain by continuous geologic substrata of impervious continental lutites. These lutites are present at relatively shallow depths (<2 m) in the hillsides of the valleys, and may temporarily create high water tables and drainage problems.

CLIMATE AND CROP WATER REQUIREMENTS

The climate of the area is Mediterranean semiarid with mean

annual maximum and minimum daily air temperatures of 19.8 and 8.9°C, respectively, mean annual precipitation of 296 mm, and mean annual reference evapotranspiration of 1200 mm. Daily climatic data (maximum, minimum, and mean temperature; precipitation; mean wind speed; maximum, minimum, and mean relative humidity; mean solar radiation) were collected with an automated weather station (Model CR21x; Campbell Scientific, Logan, UT) installed in the area. The daily reference evapotranspiration was calculated from these climatic data using the Penman–Monteith equation.

The potential evapotranspiration of each crop (PET_c) was calculated from the reference evapotranspiration and the crop coefficients derived from FAO guidelines and local agronomic information about the duration of the different phases of crop growth. The net irrigation requirement (NIR) during the irrigation period was computed as the difference between PET_c and the effective precipitation, which was estimated to be 75% of precipitation. Water samples from three rainfall events distributed throughout the study period were collected and nitrate concentrations were determined by ion chromatography.

IRRIGATION

Irrigation development in Watersheds A and B started in 1992 and was completed by 1996. Before the transformation to irrigated land the area was dryland dedicated to a winter cereals–fallow rotation. The high-quality water (electrical conductivity [EC] < 0.4 dS m^{-1}, sodium adsorption ratio [SAR] < 2) from the Monegros Canal was pumped at night to elevated reservoirs and supplied by gravity to each turnout at the demand of farmers. Watershed A had 45 irrigated fields served by 35 irrigation turncuts, which delivered water to automated solid-set sprinklers and centre pivots covering 73 and 16% of the 494-ha irrigated land, respectively.

The rest of the area is irrigated by means of big guns (3%), hand-moved sprinklers (7%), and linear-move systems (1%). Watershed B had 33 irrigated fields served by 27 irrigation turnouts, which delivered water to automated solid-set sprinklers and pivots covering 43 and 57%, respectively, of the 470 ha of irrigated land.

The volumes of irrigation water diverted by each turnout were recorded weekly by reading their corresponding water meters. From these values and the areas irrigated by each turnout, the depths (mm) of irrigation water were calculated. The nitrate concentration in the irrigation water was determined by ion chromatography monthly during the irrigation season because historical data indicated that nitrate concentration was very low (<1 mg L^{-1} NO_3–N) with low variability. Irrigation efficiency was characterized by the irrigation performance index (IPI), defined as the NIR expressed as a percentage of the volume of irrigation water (V_{iw}) delivered to the crops (IPI = NIR/V_{iw} x 100).

Irrigation performance index values less than 100% indicate overirrigation and below this threshold the IPI index is similar to the irrigation efficiency index. Values greater than 100% indicate deficit irrigation and then the IPI index is not equivalent to the irrigation efficiency. Irrigation performance index values between 85 and 115% indicate high-quality irrigation management.

NITROGEN FERTILIZATION

The crops grown in each field were recorded during the study period. A survey was conducted to determine the main characteristics of the N management by producers in 1997 for Watershed A and in 1998 for Watershed B. Farmers were asked about their methods of N application (i.e., through the irrigation water or directly to the soil) as well as the types, timing, and amount of fertilizers and manures applied.

Farmers were also asked for planting and harvesting dates, irrigation management, and the crop yields obtained. The surveyed area covered 73% (maize), 45% (alfalfa), and 100% (sunflower [*Helianthus annuus* L.] and winter cereals) of the total irrigated area in Watershed A, and 52% (maize), 83% (alfalfa), 38% (winter cereals), and 100% (sunflower, pea [*Pisum sativum* L.], and bean [*Phaseolus vulgaris* L.]) of the total irrigated area in Watershed B.

The nitrogen applied as fertilizer or manure was calculated from the survey data. The N content of the manures was taken from the literature, and we assumed that 50% of the N in the manure was available to the crop during the first growing season. We did not consider the N available from the manure applied in previous years.

The N applied in the nonsurveyed fields was estimated from the mean N application rates obtained in the corresponding surveyed crops. In October 1998, 10 maize fields, which represented 67% of the total maize area in Watershed A, were sampled in four random locations per field, two weeks after the black-layer stage. The samples consisted of the ear and the stalk portion between 0.15 and 0.35 m above the soil surface of the maize plants present in two adjacent, 1-m-long rows. The ears and the stalks were oven-dried at 60°C, and the grain yield and Kjeldhal N were determined.

A sample of 2 g of dry stalk was extracted with 10 mL of deionized water and the NO_3–N was determined colorimetrically with a continuous flow autoanalyzer (Anasol; International Controlled Atmosphere, Instrument Division, Tonbridge, UK). Four 2.5-cm-diameter soil cores were collected in each of the four locations at depths of 0 to 0.3, 0.3 to 0.6, and 0.6 to 0.9 m. A composite sample of the four soil cores at each depth was obtained.

The NO_3–N content was determined by ion chromatography after extraction of the air-dry soil samples with a saturated $CaSO_4$ solution. A bulk density of 1.4 Mg m^{-3} was used to estimate the soil NO_3–N content in kg ha^{-1}. Additional specific details about the N and crop management in these maize fields were obtained from the farmers. For the rest of crops of Watershed A (i.e., alfalfa, sunflower, and winter cereals) the mean N rates obtained in the 1997 survey were used in 1998.

DRAINAGE

A mechanical water-level recorder (Model OSK 15200-MV; Eijkelkamp Agrisearch Equipment, Giesbeek, the Netherlands) was used to monitor flow rate through a calibrated broad-crested weir at the Watershed A gauging station, whereas an electronic water level recorder (Oche, Zaragoza, Spain) monitored flow through a calibrated Parshall flume at the Watershed B gauging station. The average daily heights of water were computed from the recorded instantaneous values and converted into average daily flow rates with the appropriate calibration curves.

Instantaneous drainage water samples (0.25 L in volume) were taken every two days with automatic water samplers (Model 2900; Isco, Lincoln, NE) installed in the Watershed A and B gauging stations and the NO_3–N concentrations were determined by ion chromatography. The daily and weekly masses of NO_3–N exported from each watershed were calculated from the volumes of drainage and the NO_3–N concentrations.

The PET_c of crops and the NIR were substantially lower in 1997 than in 1998. In both years, alfalfa had the highest PET_c, followed by maize, sunflower, and winter cereals. The precipitation depths were 314 mm (well above the average value in the area) in the 1997 irrigation period and 196 mm (average value in the area) in the 1998 irrigation period. The irrigation depths (V_{iw}) applied in 1998 to maize and alfalfa were similar in Watersheds A and B and much higher than those in Watershed A during 1997.

The seasonal IPI values for maize and alfalfa were within 100 ± 15% in both irrigation periods and watersheds, whereas the rest of crops, except sunflower in 1997, were deficit-irrigated. When all the crops were computed together, the seasonal mean IPI values varied between 100 and 92%, indicating that the average irrigation management in both watersheds was excellent.

Nevertheless, the IPI values computed at the irrigation turnout level were rather variable with coefficients of variation of the seasonal mean IPI values of 25 and 36% for Watershed A in 1997 and 1998, respectively, and 29% for Watershed B in 1998, suggesting that water management differed among farmers. Thus, overirrigation (IPI < 85%) was attained in 17 and 27% for Watershed A in 1997 and 1998, respectively, and 44% for Watershed B in 1998 of the total irrigated areas.

A trend in IPI during the irrigation period was also evident, so that the IPI values tended to increase as the irrigation season progressed. Overirrigation was most common in April, mainly due to the maize postplanting irrigations given to promote maize emergence in these crusting-susceptible soils. Thus, the survey of farmer practices revealed that they irrigated maize three to five times postplanting, with an average depth of 22 mm per irrigation.

High frequency irrigation is effective against soil crusting, but taking into account that the irrigation systems allow for smaller irrigations, the irrigation depths should be minimized during this period to prevent deep percolation of water.

The analysis of the relationships between the weekly IPI and the weekly PET_c, effective precipitation, and V_{iw}, the three variables included in its calculation, indicated that the only significant ($P < 0.001$) correlation in Watersheds A and B were the negative correlations between IPI and effective precipitation, suggesting that precipitation was not adequately accounted for in the scheduling of irrigation. Presumably, the farmers hesitated to modify their computerized irrigation programming on the basis of the precipitation events.

These results show that the overall irrigation management performance in Watersheds A and B was within acceptable limits, although precipitation depths should be properly incorporated in the irrigation scheduling, the volumes of irrigation given to maize in April to promote its emergence should be minimized, and the irrigation depths given to sunflower, winter cereals, peas, and beans should be increased to match their net irrigation requirements.

NITROGEN FERTILIZATION

Maize was the most widely grown crop in the study areas, followed by alfalfa. As compared with long-term average values in nearby areas, the mean yields of maize (11–14 Mg ha^{-1}) and alfalfa (15–18 Mg ha^{-1}) were in the high range, while those for sunflower (2 Mg ha^{-1}) and winter cereals (3.5–5 Mg ha^{-1}) were in the low range. As previously indicated, the low yields of winter cereals and sunflower were probably a consequence of the insufficient irrigation, which did not meet the NIR of these crops.

Maize received 318 to 320 kg ha^{-1} of fertilizer N, similarly in both watersheds and years. These rates were moderate compared with those applied in other Spanish irrigated areas. The average available N for the crop, estimated as the fertilizer N plus 50% of the manure N, was 350 kg ha^{-1}, and the average maize yield was 12 Mg ha^{-1}. Thus, the average N application rate per unit maize grain yield was 29 kg N Mg^{-1}, similar to the recommended value of 28 to 30 kg N Mg^{-1}. However, this recommended rate was determined in flood-irrigated experiments, where irrigation efficiency is usually lower. The preplant fertilizer N applied to maize ranged between 47 and 32% of the total fertilizer N, and the manure applied at preplanting covered 53 to 60% (Watershed A) and 27% (Watershed B) of the maize growing area, with mean N rates of approximately 50 kg ha^{-1} (Watershed A) and 74 kg ha^{-1} (Watershed B).

In Watershed A, approximately 70% of the total postplant N applied to maize was supplied through the irrigation systems in one to three applications (from the six-leaf to the tassel-emergence stage) of a 32% N urea–nitrate–ammonium (50–25–25) solution, and the rest was supplied as urea (46% N) mechanically broadcasted over the soil at the maize six-leaf stage. In contrast, only 40% of the total postplant N applied to maize in Watershed B was supplied through the irrigation systems.

The rates of N fertilizer applied to alfalfa, peas, and beans were low (50–55 kg ha^{-1}). Sunflower received moderate N rates (62–100 kg ha^{-1}) with 54 to 70% applied at preplanting and the rest in a postplant application through the irrigation systems. Winter cereals received approximately 150 kg N ha^{-1} with 26 to 34% applied at preplanting and the rest as ammonium nitrate (33.5% N) or urea applied mechanically.

The results of the surveyed (Watershed A in 1997 and Watershed B in 1998) and measured (Watershed A in 1998) maize fields indicate that grain yields increased with increases in the available N up to approximately 400 kg ha^{-1}, and then they tended to decrease. However, only 6 out of the 31 maize fields surveyed had available N values higher than 400 kg ha^{-1}.

Maize grain yield was independent of the proportion of the total

N applied at preplanting, although yields higher than approximately 12 Mg ha^{-1} were only found when less than about 45% of the N was applied before planting. Higher proportions of preplant N were ineffective in promoting maize yields and could lead to potential losses of N below the root zone.

The results of the 10 maize fields sampled in October 1998 in Watershed A suggest that, within the measured ranges, grain N was independent of total and postplant fertilizer N· The stalk N and the stalk NO_3–N tended to increase as the N applied increased just above thresholds for total and postplant N of 350 kg ha^{-1} and 300 kg ha^{-1}, respectively. Stalk NO_3–N contents measured in 70% of the maize fields were above the 0.07 to 0.2% range proposed by Binford et al. (1992) as indicative of N overfertilization, and the stalk N contents measured in all the maize fields were higher than the value of 0.43% proposed by Binford et al. (1990) as indicative of excess N. However, these threshold values were obtained under more humid climatic conditions and without irrigation.

According to Villar (1999), a stalk N threshold value of 0.5% would be more appropriate for the climatic and cropping conditions of this study. Thus, considering the stalk N and stalk NO_3–N found and the thresholds established by Villar (1999) and Binford et al. (1992), respectively, 60 to 70% of the sampled fields were overfertilized.

The NO_3–N content of the soils sampled at maize harvest ranged between 42 and 284 kg ha^{-1}, with a mean value of 117 kg ha^{-1}. With some exceptions, soil NO_3–N tended to increase with increasing applications of total and postplant N. However, the soil NO_3–N content at maize harvest is not only determined by the balance of N inputs and N uptake, but also by the N losses during the growing season that are mostly related to irrigation management. Thus, the field that received the highest rate of fertilizer N had a low value of 105 kg soil NO_3–N ha^{-1} due to nitrate leaching derived from the shallow-depth (<0.35 m) soil in this field and the high volume of irrigation water applied, which was 33% above the maize NIR.

The N and NO_3–N content in the basal portion of the maize stalk and the residual soil NO_3–N content at maize harvesting found in our work indicate that the N rates applied to maize could be reduced, thereby decreasing the potential for nitrate losses in drainage waters. In central Spain, Diez et al. (2000) found similar yields when applying 150 kg N ha^{-1}.

Our results also show that the timing of applications could be improved by decreasing the proportion of total N applied at preplanting (high maize yields were only obtained when preplant N was less than 45% of total N), which could be easily accomplished by increasing the N applications through the irrigation system.

In general, drainage flow rates measured at the Watershed A and B gauging stations were lowest during the winter months and immediately before the commencement of the irrigation period in March. Drainage increased during the irrigation season and decreased again at the end of the irrigation period in September. Obviously, this trend was due to the irrigation return flows resulting from excess water applied to some fields. However, precipitation also affected drainage flows, as shown by the higher flows measured at the Watershed A gauging station in 1997 (irrigated-season precipitation = 314 mm) than in 1998 (irrigated-season precipitation = 196 mm).

The high-quality irrigation management in both study areas resulted in relatively low drainage flows, which represented 5 to 11% of the irrigation + precipitation water entering in Watershed A and 18% of the irrigation + precipitation water entering in Watershed B.

The patterns of flow measured at the Watershed A and B gauging stations were similar, but higher in the Watershed B gauging station than in the Watershed A gauging station. Thus, for the 1998 hydrological year, October 1997 through September 1998, the volumes of drainage per unit irrigated area were 194 mm in Watershed B but only 48 mm in Watershed A.

The higher drainage in Watershed B was attributed to (i) the seepage of the Monegros Canal, that accounted for 33 mm during the irrigation period, (ii) the contribution from precipitation falling on the larger area of nonirrigated land within this watershed, and (iii) the larger overirrigated area in Watershed B (44% of the total irrigated area in 1998) than in Watershed A (27% of the total irrigated area in 1998).

The NO_3–N concentrations in the drainage waters ranged between 7 and 55 mg L^{-1} (Watershed A gauging station) and between 9 and 42 mg L^{-1} (Watershed B gauging station), and 98% of the water samples had values above 10 mg L^{-1}. In general, the weekly NO_3–N concentrations were higher at the beginning (April–May) and end (August–September) of the irrigation periods, and decreased during June and July.

The higher concentrations at the beginning of the irrigation periods were attributed to the preplanting applications of N to maize in a period of low irrigation efficiencies, and to the postplant applications of N to winter cereals, whereas the lower concentrations in June and July were related to the period of maximum N uptake by maize, coupled to IPI values higher than 100% in July.

In contrast, Isidoro (1999) found higher drainage NO_3–N

concentrations in June and July than in April and May in La Violada irrigation district. The opposite behaviour in these two nearby areas was attributed to differences in nitrogen and irrigation water management. All the maize postplant N was applied to the soil in June in La Violada, as compared with the split applications through the irrigation water in Watersheds A and B.

Moreover, in June and July the IPI values in the La Violada level-basin irrigated district were lower than 50%, as compared with values between 70 and 110% in Watersheds A and B. The mean NO_3–N concentrations during the irrigated and nonirrigated seasons were similar (not significantly different at $P < 0.05$) in both watersheds, ranging between 26 and 29 mg L^{-1} at the Watershed A gauging station and between 23 and 28 mg L^{-1} at the Watershed B gauging station.

The lower irrigated-season NO_3–N concentration at the Watershed B gauging station (23 mg L^{-1}) than at the Watershed A gauging station (29 mg L^{-1}) may be a result of the low-nitrate Monegros Canal seepage water collected at the Watershed B gauging station.

The NO_3–N concentrations of the drainage water were similar to those found in other studies and moderately higher than the 20 mg L^{-1} value suggested by Keeney (1982) as the lowest achievable concentration in the drainage waters of many irrigated fields under good agronomic practices and profitable crop productions. However, these concentrations are double the threshold human consumption NO_3–N concentrations established by the USEPA (10 mg L^{-1}) and the European Union (11.3 mg L^{-1}).

WATER BALANCE

Deep percolation was not measured, although it may be approximated by the input - output difference if steady state conditions are assumed. This assumption may be acceptable for an annual period, but not for just the irrigated or the nonirrigated periods. The water balance in Watershed A indicated that deep percolation was negligible and that the difference between inputs and outputs was less than 4% of the total input or output water.

The negative value obtained in Watershed B suggests that there was not deep percolation and that the outputs were overestimated and/or the inputs were underestimated. Thus, in Watershed B the actual evapotranspiration was probably lower than the calculated PET_c, since this was estimated assuming maximum potential crop yields, and the peas–beans and sunflower crops were underirrigated.

Also, a nonmeasured ground water input to the Watershed B study area could result from precipitation in its large dry-land watershed, which could explain its negative water balance. In any case, the difference between the inputs and outputs in Watershed B was less than 10% of the total input or output water, indicating that the closure of the water balance was satisfactory.

The better closure of the water balance in Watershed A than in Watershed B during the study period, but also when comparing the same time period (October 1997–September 1998), suggest that results from Watershed A were more reliable.

The estimated mass of N fertilizer and manure applied to Watersheds A and B varied between 166 and 221 kg ha^{-1} in the irrigated seasons and between 3 and 32 kg ha^{-1} in the nonirrigated seasons, representing more than 93% (irrigated season) and 57% (nonirrigated season) of the total imported mass of N. This is an expected result due to the low average NO_3–N concentrations measured in the irrigation (1.14 mg L^{-1}) and precipitation (1.94 mg L^{-1}) waters.

The weekly NO_3–N loads paralleled the weekly drainage flow rates in both monitoring stations, and they were linearly correlated ($P < 0.001$) so that each millimeter of drainage produced losses of 0.27 (Watershed A) and 0.25 (Watershed B) kg NO_3–N ha^{-1}. Thus, as found in other studies, the volume of drainage basically determined the exported mass of nitrates.

The mass of NO_3–N exported by Watershed A drainage waters varied between 19 and 9 kg ha^{-1} in the irrigated periods and between 5 and 4 kg ha^{-1} in the nonirrigated periods. The mass of NO_3–N exported by the Watershed B drainage waters was higher than that in Watershed A and similar (i.e., approximately 25 kg ha^{-1}) in the irrigated and nonirrigated periods. As previously indicated, the higher NO_3–N loads in Watershed B were mainly attributed to the higher drainage flows derived from the precipitation falling in its larger watershed.

Other reasons for the higher NO_3–N loads in Watershed B could be its larger relative area cropped with maize (57% and 42–54% of the irrigated areas in Watersheds B and A, respectively), which was the crop with the highest N fertilizer rates, and also the lower postplant application of N to maize through the irrigation system (40 and 70% of the postplant N in Watersheds B and A, respectively). The appropriate splitting of N in the irrigation water has been proposed as an efficient and low-cost practice to reduce nitrate leaching losses in areas of low precipitation.

The higher NO_3–N load in Watershed A during the irrigation period of 1997 compared with the irrigation period of 1998 was attributed to the above-average precipitation of 314 mm in 1997, as compared with the precipitation of 196 mm in 1997, which was similar to the historical average. Since precipitation events are for the most part unpredictable, it should be recognized that some nitrate losses are unavoidable, even under a farmer's best management practices.

The mass of exported N expressed in per cent of the applied fertilizer plus manure N was 5 to 9% for irrigation periods and 15 to 25% for nonirrigation periods in Watershed A, and 11% for the irrigation period and 733% for the nonirrigation period in Watershed B. Considering the entire study periods for each watershed, NO_3–N loads represented 8.5% of the applied fertilizer plus manure N in Watershed A and 21.8% in Watershed B.

Since closure of the water balance was better in Watershed A than in Watershed B and the hydrological limits of Watershed B were not restricted to the studied area, the 8.5% figure founa in Watershed A can be considered as more representative of the NO_3–N load losses derived from irrigated agriculture in this area.

The NO_3–N loads measured in the drainage waters were, in general, lower than those reported in other watersheds of the Ebro River basin with similar crops and climatic conditions but surface-irrigated. Basso (1994) reported in a three-year study mean annual NO_3–N losses of 35 to 59 kg ha^{-1} in the Bardenas I irrigation district, which represented 16 to 30% of the applied fertilizer N. Causapé et al. (2002) found annual NO_3–N losses of 98 and 195 kg ha^{-1}, representing 44 and 56% of the applied fertilizer N, in two watersheds of the Bardenas I district. Finally, Isidoro (1999) reported in a two-year study in the La Violada irrigation district (Monegros I) mean annual NO_3–N losses of 68 kg ha^{-1} or 23% of the applied fertilizer N.

The drainage NO_3–N losses, particularly those found in Watershed A that can be considered as more representative for our study area, were in the low range of those reported in other studies around the world. In the semiarid conditions of west-central Nebraska (394 mm of precipitation from April to September), Klocke et al. (1999) found under sprinkler irrigation mean annual losses of 52 kg NO_3–N ha^{-1} (27% of applied fertilizer N) for continuous maize and 91 kg NO_3–N ha^{-1} (105% of applied fertilizer N) for a maize–soybean [*Glycine max* (L.) Merr.] rotation.

The Management Systems Evaluation Area (MSEA) results from Iowa (nonirrigated maize and soybean) found annual losses of 4 to 66 kg NO_3–N ha^{-1}, representing 6 to 115% of the applied fertilizer N. Pratt (1984) found in various irrigated areas of California annual average losses of 82 kg NO_3–N ha^{-1} (50% of the applied fertilizer N).

In summary, despite the relatively high drainage NO_3–N concentrations found in our work, the low drainage discharges derived from high-quality irrigation management resulted in NO_3–N loads that were in the low range of those measured in other irrigated areas. Under the semiarid climatic conditions of the Ebro River basin, most water for crops is provided through irrigation. Consequently, irrigation management greatly influences nitrate losses in drainage waters. The high quality of irrigation water in the Monegros irrigation area allows for negligible leaching fractions.

Our work indicates that the high quality of irrigation management (IPI within 100 ± 15) attained with sprinkler irrigation, coupled to the postplanting splitting of fertilizer N through the irrigation systems, resulted in relatively low annual drainage NO_3–N losses (8.5% of the applied N for the more representative and well-managed watershed). Nevertheless, there is room for improvement within the studied watersheds, since

- A substantial part of the plots (17–44% of the total irrigated area, depending on years and watersheds) was overirrigated,
- Farmers did not account properly for precipitation on the scheduling of irrigation,
- Postplanting irrigation depths given to promote maize emergence could be decreased,
- A large proportion (60–70%) of the sampled maize fields were overfertilized, and
- A larger fraction of fertilizer N could be split through its application with the irrigation water rather than applying it as preplant fertilizer.

Taking into account these considerations, nitrate leaching losses in the drainage waters of these irrigated watersheds could be further decreased. Although nitrogen pollution regulations are generally based on maximum allowable concentrations (MAC), the mass of nitrogen exported per unit irrigated area is important for controlling the nitrogen contamination in the receiving water bodies. Thus, even though drainage nitrate concentrations in the study areas are above the MAC, their nitrogen loads can be considered low, allowing for the attainment of a sensible compromise.

Chapter 11

Drainage

A drainage basin is an extent of land where water from rain or snow melt drains downhill into a body of water, such as a river, lake, reservoir, estuary, wetland, sea or ocean. The drainage basin includes both the streams and rivers that convey the water as well as the land surfaces from which water drains into those channels, and is separated from adjacent basins by a drainage divide.

The drainage basin acts like a funnel, collecting all the water within the area covered by the basin and channelling it into a waterway. Each drainage basin is separated topographically from adjacent basins by a geographical barrier such as a ridge, hill or mountain, which is known as a water divide.

Other terms that are used to describe a drainage basin are *catchment*, catchment area, catchment basin, drainage area, river basin, water basin and watershed. In the technical sense, a watershed refers to a divide that separates one drainage area from another drainage area. However, in the United States and Canada, the term is often used to mean a drainage basin or catchment area itself. Watersheds drain into other watersheds in a hierarchical form, larger ones breaking into smaller ones or sub-watersheds with the topography determining where the water flows. Understanding geomorphology is essential in understanding how watersheds interconnect.

OCEAN BASINS

There are numerous drainage basins throughout the world in all sorts of places. The following is a list of some of the major ones:

·The Atlantic Ocean drains approximately 47% of all land in the world. In North America, it directly drains the Saint Lawrence River and Great Lakes basins, the Eastern Seaboard of the United States, the Canadian Maritimes, and most of Newfoundland and Labrador. It also directly drains nearly all of South America east of the Andes, most of Western and Central Europe, and the greatest portion of western Sub-Saharan Africa.

The three major mediterranean seas of the world also flow to the Atlantic:

- The American Mediterranean Sea (the Caribbean Sea and Gulf of Mexico) basin includes most of the American interior between the Appalachian and Rocky Mountain ranges, a small part of the Canadian provinces of Alberta and Saskatchewan, eastern Central America, the islands of the Caribbean and the Gulf, and a small part of northern South America.
- The European Mediterranean Sea basin includes much of northern Africa, east-central Africa (through the Nile), southern, central, and eastern Europe, Turkey, and the coastal areas of Israel, Lebanon, and Syria.
- The Arctic Ocean basin drains most of Western and Northern Canada east of the Continental Divide, the north shore of Alaska and parts of North Dakota, South Dakota, Minnesota, and Montana in the United States, the north shore of the Scandinavian peninsula in Europe, and much of central and northern Russia.
- The Pacific Ocean drains just over 13% of the land in the world. Its basin includes much of China, southeastern Russia, Japan, Korea, most of Indonesia and Malaysia, the Philippines, all of the Pacific Islands, the northeast coast of Australia, and the western parts of Canada, the United States (including most of Alaska), Central America, and South America.
- The Indian Ocean drains around 13% of the Earth's land. It drains the eastern coast of Africa, the coasts of the Red Sea and the Persian Gulf, the Indian subcontinent, Burma, and most of Australia.
- The Southern Ocean drains Antarctica. Antarctica comprises approximately eight per cent of the Earth's land.

Largest River Basins

The three largest river basins (by area), from largest to smallest, are the Amazon basin, the Congo basin, and the Mississippi basin. The three rivers that drain the most water, from most to least, are the Amazon, Congo, and Ganges Rivers.

ENDORHEIC DRAINAGE BASINS

Endorheic drainage basins are inland basins that do not drain into an ocean; around 18% of all land drains to endorheic lakes or seas. The largest of these consists of much of the interior of Asia, and drains into the Caspian Sea and the Aral Sea. Other basins include the Great

Basin in the United States, much of the Sahara Desert, the watershed of the Okavango River (Kalahari Basin), highlands near the African Great Lakes, the interiors of Australia and the Arabian Peninsula, and parts in Mexico and the Andes.

In endorheic bodies of standing water where evaporation is the primary means of water loss, the water is typically more saline than the oceans. An extreme example is the Dead Sea. An endorheic basin (from the Greek: —íäïí, *éndon,* "within" and ååÖí, *rheîn,* "to flow"; also *terminal* or *closed basin*) is a closed drainage basin that retains water and allows no outflow to other bodies of water such as rivers or oceans.

Normally the water accruing in drainage basins flows out through surface rivers or by underground diffusion through permeable rock to the oceans. However, in an endorheic basin, rain (or other precipitation) that falls within it does not flow out but may only leave the drainage system by evaporation and seepage. The bottom of such a basin is typically occupied by a salt lake or salt pan. Endorheic basins are also called internal drainage systems.

Endorheic regions, in contrast to exorheic regions which flow to the ocean in geologically defined patterns, are closed hydrologic systems. Their surface waters drain to inland terminal locations where the water evaporates or seeps into the ground, having no access to discharge into the sea. Endorheic water bodies include some of the largest lakes in the world, such as the Aral Sea and the Caspian Sea, the world's largest saline body of water cut off from the ocean.

ENDORHEIC LAKES

Endorheic lakes are bodies of water that do not flow into the sea. Most of the water falling on earth finds its way to the oceans through a network of rivers, lakes and wetlands or through evaporation. However, there is a class of water bodies that are located in closed or endorheic watersheds where the topography prevents their drainage to the oceans. These endorheic watersheds (containing water in rivers or lakes that form a balance of surface inflows, evaporation and seepage) are often called terminal lakes or sink lakes.

Endorheic lakes are usually in the interior of a body mass, far from an ocean. Their watersheds are often confined by natural geologic land formations such as a mountain range, cutting off water access to the ocean. The inland water flow into dry watersheds where the water evaporates, leaving a high concentration of minerals and other inflow erosion products.

Over time this input of erosion products can cause the Endorheic Lake to become relatively saline. Since the main outflow pathways of these lakes are chiefly through evaporation and seepage, endorheic lakes are usually more sensitive to environmental pollutants inputs than water bodies that have access to oceans.

Endorheic regions can occur in any climate but are most commonly found in hot desert locations. In areas where rainfall is higher, riparian erosion will generally carve drainage channels (particularly in times of flood), or cause the water level in the terminal lake to rise until it finds an outlet, breaking the enclosed endorheic hydrological system's geographical barrier and opening it to the surrounding terrain. The Black Sea was likely such a lake, having once been an independent hydrological system before the Mediterranean Sea broke through the terrain separating the two.

Endorheic regions tend to be far inland with their boundaries defined by mountains or other geological features that block their access to oceans. Since the inflowing water can evacuate only through seepage or evaporation, dried minerals or other products collect in the basin, eventually making the water saline and also making the basin vulnerable to pollution. Continents vary in their concentration of endorheic regions due to conditions of geography and climate. Australia has the highest percentage of endorheic regions at 21 per cent while North America has the least at 5 per cent. Approximately 18 per cent of the earth's land drains to endorheic lakes or seas, the largest of these land areas being the interior of Asia.

In deserts, water inflow is low and loss to solar evaporation high, drastically reducing the formation of complete drainage systems. Closed water flow areas often lead to the concentration of salts and other minerals in the basin. Minerals leached from the surrounding rocks are deposited in the basin, and left behind when the water evaporates. Thus endorheic basins often contain extensive salt pans (also called salt flats, salt lakes, alkali flats or playas). These areas tend to be large, flat hardened surfaces and are sometimes used for aviation runways or land speed record attempts, because of their extensive areas of perfectly level terrain.

Both permanent and seasonal endorheic lakes can form in endorheic basins. Some endorheic basins are essentially stable, climate change having reduced precipitation to the degree that a lake no longer forms. Even most permanent endorheic lakes change size and shape dramatically over time, often becoming much smaller or breaking into several smaller parts during the dry season.

As humans have expanded into previously uninhabitable desert areas, the river systems that feed many endorheic lakes have been altered by the construction of dams and aqueducts. As a result many endorheic lakes in developed or developing countries have contracted dramatically, resulting in increased salinity, higher concentrations of pollutants, and the disruption of ecosystems.

NOTABLE ENDORHEIC BASINS AND LAKES

Antarctica

Endorheic lakes in Antarctica are located in the McMurdo Dry Valleys, Victoria Land, Antarctica, the largest ice-free area in Antarctica.

- Don Juan Pond in Wright Valley is fed by groundwater from a rock glacier and remains unfrozen throughout the year.
- Lake Vanda in Wright Valley has a perennial ice cover, the edges of which melt in the summer allowing flow from the longest river in Antarctica, the Onyx River. The lake is over 70 m deep and is hypersaline.
- Lake Bonney is in Taylor Valley and has a perennial ice over and two lobes separated by the Bonney Riegel. The lake is fed by glacial melt and discharge from Blood Falls. Its unique glacial history has resulted in a hypersaline brine in the bottom waters and fresh water at the surface.
- Lake Hoare, in Taylor Valley, is the freshest of the Dry Valley lakes receiving its melt almost exclusively from the Canada Glacier. The lake has an ice cover and forms a moat during the Austral summer.
- Lake Fryxell, in adjacent to the Ross Sea in Taylor Valley. The lake has an ice cover and receives its water from numerous glacial meltwater streams for approximately 6 weeks out of the year. Its salinity increases with depth.

Asia

Much of western and Central Asia is a single, giant inland basin. It contains several lakes, including:

- The Central Asian Internal Drainage Basin, the largest of the three major basins covering Mongolia.
- The Caspian Sea, the largest lake on Earth. In fact, a large part of Eastern Europe drained by the Volga River also belongs to its basin.
- The Aral Sea, whose tributary rivers have been diverted, leading to a dramatic shrinkage of the lake. The resulting ecological disaster has brought the plight faced by internal drainage basins to public attention.
- Lake Balkhash (Kazakhstan)

- Lop Nur Basin, in the southeastern portion of Xinjiang Uygur Autonomous Region in northwestern China.
- Issyk-Kul, Son-Kul and Chatyr-Kul lakes in Kyrgyzstan.
- Sistan Basin covering areas of Iran and Afghanistan.
- Tarim Basin in Xinjiang Uyghur Autonomous Region.
- Uvs Nuur basin, Mongolia, Tuvan Republic of Russia.
- The Dead Sea, the lowest surface point on Earth and one of its saltiest bodies of water, lies between Israel and Jordan.
- Sambhar Lake in Rajasthan, north-western India, is also the terminal point of an endorheic basin.
- Lake Van in Turkey is one of the world's largest endorheic lakes.
- Solar Lake, Sinai, near the Israeli-Egypt border.

IMPORTANCE OF DRAINAGE BASINS

GEOPOLITICAL BOUNDARIES

Drainage basins have been historically important for determining territorial boundaries, particularly in regions where trade by water has been important. For example, the English crown gave the Hudson's Bay Company a monopoly on the fur trade in the entire Hudson Bay watershed, an area called Rupert's Land. Today, bioregional democracy can include agreements of states in a particular drainage basin to defend it. One example of this is the Great Lakes Commission.

Hydrology

In hydrology, the drainage basin is a logical unit of focus for studying the movement of water within the hydrological cycle, because the majority of water that discharges from the basin outlet originated as precipitation falling on the basin. A portion of the water that enters the groundwater system beneath the drainage basin may flow towards the outlet of another drainage basin because groundwater flow directions do not always match those of their overlying drainage network. Measurement of the discharge of water from a basin may be made by a stream gauge located at the basin's outlet.

Rain gauge data is used to measure total precipitation over a drainage basin, and there are different ways to interpret that data. If the gauges are many and evenly distributed over an area of uniform precipitation, using the arithmetic mean method will give good results. In the Thiessen polygon method, the watershed is divided into polygons with the rain gauge in the middle of each polygon assumed to be representative for the rainfall on the area of land included in its polygon.

These polygons are made by drawing lines between gauges, then making perpendicular bisectors of those lines form the polygons. The isohyetal method involves contours of equal precipitation are drawn over the gauges on a map. Calculating the area between these curves and adding up the volume of water is time consuming.

Ecology

Drainage basins are important elements to consider also in ecology. As water flows over the ground and along rivers it can pick up nutrients, sediment, and pollutants. Like the water, they get transported towards the outlet of the basin, and can affect the ecological processes along the way as well as in the receiving water source.

Modern usage of artificial fertilizers, containing nitrogen, phosphorus, and potassium, has affected the mouths of watersheds. The minerals will be carried by the watershed to the mouth and accumulate there, disturbing the natural mineral balance. This can cause eutrophication where plant growth is accelerated by the additional material.

Resource Management

Because drainage basins are coherent entities in a hydrological sense, it has become common to manage water resources on the basis of individual basins. In the U.S. state of Minnesota, governmental entities that perform this function are called watershed districts. In New Zealand, they are called catchment boards.

Comparable community groups based in Ontario, Canada, are called conservation authorities. In North America this function is referred to as watershed management. In Brazil, the National Policy of Water Resources, regulated by Act n° 9.433 of 1997, establishes the drainage basin as territorial division of Brazilian water management.

Watershed management is the process of creating and implementing plans, programs, and projects to sustain and enhance watershed functions that affect the plant, animal, and human communities within a watershed boundary. Features of a watershed that agencies seek to manage include water supply, water quality, drainage, stormwater runoff, water rights, and the overall planning and utilization of watersheds.

Landowners, land use agencies, stormwater management experts, environmental specialists, water use purveyors and communities all play an integral part in the management of a watershed.

Sources of Pollution

In an agricultural landscape, common contributors to water pollution are nutrients and sediment which typically enter stream systems after rainfall washes them off poorly managed agricultural fields, called surface runoff, or flushes them out of the soil through leaching. These types of pollutants are considered nonpoint source pollution because the exact point where the pollutant originated cannot be identified. Point source pollution originates a specific point of contamination such as if a manure containment structure fails and its contents enter the drainage system.

In urban landscapes, issues of soil loss through erosion, from construction sites for example, and nutrient enrichment from lawn fertilizers exist. Point source pollution, such as effluent from wastewater treatment plants and other industries play a much larger role in this setting. Also, the greatly increased area of impervious surfaces, such as concrete, combined with modern storm drainage systems, allows for water and the contaminants that it can carry with it to exit the urban landscape quickly and end up in the nearest stream.

Controlling Pollution

In agricultural systems, common practices include the use of buffer strips, grassed waterways, the reestablishment of wetlands, and forms of sustainable agriculture practices such as conservation tillage, crop rotation and intercropping. After certain practices are installed, it is important to continually monitor these systems to ensure that they are working properly in terms of improving environmental quality.

In urban settings, managing areas to prevent soil loss and control stormwater flow are a few of the areas that receive attention. A few practices that are used to manage stormwater before it reaches a channel are retention ponds, filtering systems and wetlands.

It is important that stormwater is given an opportunity to infiltrate so that the soil and vegetation can act as a "filter" before the water reaches nearby streams or lakes. In the case of soil erosion prevention, a few common practices include the use of silt fences, landscape fabric with grass seed and hydroseeding. The main objective in all cases is to slow water movement to prevent soil transport.

The Roles of Federal Agencies

In the United States, the Natural Resources Conservation Service

(NRCS) and the United States Environmental Protection Agency (EPA) are responsible for work on the federal level. The NRCS is typically involved with the planning and continued monitoring of environmental improvement projects, while the EPA is generally responsible for compliance of several environmental laws such as the Clean Water Act. Assistance with watershed protection is also provided on a state level through Soil and Water conservation districts and other state-operated departments (e.g., departments of natural resources, departments of agriculture).

Beyond governmental support, other organizations and companies exist that provide support in various manners with the goal of watershed protection in mind as well.

ENVIRONMENTAL LAW

Environmental laws often dictate the planning and actions that agencies take to manage watersheds. Some laws require that planning be done, others can be used to make a plan legally enforceable and others set out the ground rules for what can and cannot be done in development and planning. Most countries and states have their own laws regarding watershed management.

Those concerned about aquatic habitat protection have a right to participate in the laws and planning processes that affect aquatic habitats. By having a clear understanding of who to speak to and how to present the case for keeping our waterways clean a member of the public can become an effective watershed protection advocate.

Environmental law is a complex and interlocking body of statutes, common law, treaties, conventions, regulations and policies which, very broadly, operate to regulate the interaction of humanity and the rest of the biophysical or natural environment, toward the purpose of reducing or minimizing the impacts of human activity, both on the natural environment for its own sake, and on humanity itself. Environmental law draws from and is influenced by principles of environmentalism, including ecology, conservation, stewardship, responsibility and sustainability. From an economic perspective it can be understood as concerned with the prevention of present and future externalities.

Areas of concern in environmental law include air quality, water quality, global climate change, agriculture, biodiversity, species protection, pesticides and hazardous chemicals, waste management, remediation of contaminated land and brownfields, smart growth, sustainable development, impact review, and conservation, stewardship and management of public lands and natural resources. While many countries worldwide have accumulated impressive sets of environmental laws, their implementation has often been woeful. In recent years, environmental law has become seen as a critical means of promoting sustainable development (or "sustainability").

Policy concepts such as the precautionary principle, public participation, environmental justice, and the polluter pays principle have informed many environmental law reforms in this respect. There has been considerable experimentation in the search for more effective methods of environmental control beyond traditional "command-and-control" style regulation. Eco-taxes, emission trading, voluntary standards such as ISO 14000 and negotiated agreements are some of these innovations.

Environmental law is offered as an elective course in the second and third year of JD study at many American law schools. Some U.S. law schools also offer an LLM or JSD specialization in environmental law. Additionally, several law schools offer legal clinics that focus on environmental law, providing students with an opportunity to learn about environmental law in the context of real world disputes involving actual clients.

The IUCN Academy of Environmental Law is a network of some 60 law schools worldwide that specialise in the research and teaching of environmental law.

U.S. News & World Report has consistently ranked University of Oregon School of Law, Vermont Law School, Lewis & Clark Law School, Pace University School of Law, Tulane University School of Law, and Georgetown University Law Centre as among the best Environmental Law programs in the United States.

However, the ranking system for environmental law programs is veiled in secrecy and apparently does not consult lawyers for their opinions. The Deans of U.S. law schools have repeatedly issued statements warning against relying on law school ratings, which are produced by a general news magazine with commercial motives.

Catchment Factors

The catchment is the most significant factor determining the amount or likelihood of flooding.

Catchment factors are: topography, shape, size, soil type and land use (paved or roofed areas). Catchment topography and shape determine the time taken for rain to reach the river, Catchment size, soil type and development determine the amount of water to reach the river.

Topography

Topography determines the speed with which the runoff will reach a river. Clearly rain that falls in steep mountainous areas will reach the river faster than flat or gently sloping areas.

Shape

Shape will contribute to the speed with which the runoff reaches a river. A long thin catchment will take longer to drain than a circular catchment.

Size

Size will help determine the amount of water reaching the river, as the larger the catchment the greater the potential for flooding.

Soil Type

Soil type will help determine how much water reaches the river. Certain soil types such as sandy soils are very free draining and rainfall on sandy soil is likely to be absorbed by the ground. However, soils containing clay can be almost impermeable and therefore rainfall on clay soils will run off and contribute to flood volumes. After prolonged rainfall even free draining soils can become saturated, meaning that any further rainfall will reach the river rather than being absorbed by the ground.

Land use

Land use can contribute to the volume of water reaching the river, in a similar way to clay soils. For example, rainfall on roofs, pavements and roads will be collected by rivers with almost no absorption into the groundwater.

RESERVOIR

A reservoir is, most broadly, a place or hollow vessel where fluid is kept in reserve, for later use. Most often, a reservoir refers to an artificial lake, used to store water for various uses. Reservoirs are often created by building a sturdy dam, usually out of concrete, earth, rock, or a mixture across a river or stream. Once the dam is completed, the stream fills the reservoir. When a reservoir is predominantly man-made (rather than being an adaptation of a natural basin) it may be called a cistern. The term reservoir is also often used to describe underground reservoirs such as an oil or water well.

Many reservoirs have 'lake' in their names; a few natural lakes have 'reservoir' in *their* names. There are three basic types of reservoir:

the commonly-seen dam across a valley, the less-common fully-bounded reservoir and sealed and covered reservoir to store high quality material safe from contamination. In the case of drinking water these covered reservoirs are often termed *Service reservoirs*.

CONSTRUCTION AND USE

Valley Dammed Reservoir

The more common dam across a valley relies on naturally formed features to form the water tight elements. Generally, engineers look for dam sites which are narrow with a broad area upstream; the valley sides can then act as natural walls and the broad area upstream makes a large reservoir for the height.

The best place along the valley for building a dam has to be determined according to where the dam can best be tied into the valley walls and floor to form a water tight seal. If necessary, humans have to be re-housed or historic sites must be moved. For example, the temples of Abu Simbel were moved before the construction of the Aswan Dam (which created Lake Nasser from the Nile in Egypt).

At the start of construction the river must be diverted, often through a tunnel. Then the foundation is prepared. Once that is done, building of the dam can start. This may take anywhere from a few months to a few years, depending on its size and complexity. After the dam is complete, the diversion is removed or plugged, and the river fills the area upstream of the dam.

Bank-Side Reservoir

Where water is taken from a river of variable quality or quantity, it is common to construct bank-side reservoirs to store water pumped or siphoned from the river. Such reservoirs are usually built partly by excavation and partly by the construction of a complete encircling bund or embankment. Both the floor of the reservoir and the bund must have an impermeable lining or core, often made of puddled clay.

The water stored in such reservoirs may have a residence time of several months during which time normal biological processes are able to substantially reduce many contaminants and almost eliminate any turbidity. The use of bank-side reservoirs also allows a water abstraction to be closed down for extended period at times when the river is unacceptably polluted or when flow conditions are very low due to drought.

The London water supply system is one example of the use of bank-side storage for all the water taken from the River Thames and

River Lee with many large reservoirs visible along the approach to Heathrow airport.

Service Reservoir

Many service reservoirs are constructed as water towers, often as elevated structures on concrete pillars where the landscape is relatively flat. Other service reservoirs are entirely underground, especially in more hilly or mountainous country.

In the United Kingdom Thames Water has many underground reservoirs beneath London built in the 1800s by the Victorians, most of which are lined with thick layers of brick. Honour Oak Reservoir, which was completed in 1909, is the largest of this type in Europe. The roof is supported using large brick pillars and arches and the outside surface is used as a golf course.

Operation

A raw water reservoir does not simply hold water until it is needed. It is the first part of the water treatment process. The time the water is held for before it is released is known as the *retention time*. This is a design feature that allows particles and silts to settle out, as well as time for natural biological treatment using algae, bacteria and zooplankton that naturally live within the water.

Water can be released from the reservoir, generally by gravity, to be cleaned for drinking water, generate electricity, or simply maintain the downstream flow. In the event that major rainfall occurs, water can be released via a spillway to avoid over-topping and compromising the integrity of the dam. Most modern reservoirs have a specially designed draw-off tower that can discharge water from the reservoir at different levels both to access water as the reservoir draws down but also to allow water of a specific quality to be discharged into the downstream river as compensation water.

Levels

The terminology for reservoirs varies from country to country. In the United States the normal maximum level of a reservoir lake is called *full pool*, while the minimum level it can function at is *dead pool*. The water below this point is also called the dead pool, while the water in between is called the *conservation pool*. Full pool may have different levels in summer and winter, or based on the local wet and dry seasons.

Once a reservoir reaches dead pool, it is below the level at which the dam can release it downstream. At this point, the streambed beyond the dam goes nearly or completely dry, and electricity production stops.

This is also often the point at which intakes for municipal water systems begin to suck air in, and must be extended into deeper water, where stagnant water quality is much poorer. This can be done either permanently with longer pipes, or temporarily with large hoses floated on small barges, such as until a severe drought or dam repairs are over.

HYDROELECTRICITY

A hydroelectric power station consists of large turbines at the base of a dam. Water from the reservoir behind the dam is channeled through pipes and delivered to the turbines, which in turn, spin a generator to produce electricity. Generally, reservoirs behind hydroelectric dams are built specifically for electrical power generation and are not used for drinking water or irrigation.

Controlling Watercourses

Reservoirs can be used in a number of ways to control how water flows through downstream waterways.

Irrigation

Water in an irrigation reservoir is released into networks of canals mainly for use in farmlands or secondary water systems. Water in an irrigation reservoir is *generally* not used for drinking water, but in some cases is.

Flood Control

Commonly known as an "*attenuation*" or "*balancing*" reservoir, these are used to prevent flooding to lower lying lands, flood control reservoirs collect water at times of unseasonally high rainfall, then release it slowly over the course of the following weeks or months. Some of these reservoirs are constructed across the river line with the onward flow controlled by an orifice plate.

When river flow exceeds the capacity of the orifice plate water builds behind the dam but as soon as the flow rate reduces the water behind the dam slowly releases until the reservoir is empty again. In some cases such reservoirs only function a few times in a decade and the land behind the reservoir may be developed as community or recreational land.

Compensation

If a standard reservoir is built on a river which is used as a source of power, a compensation reservoir may also be built to guarantee a sufficient flow of water downstream during the working hours of the water-powered industries.

Canals

Where a natural watercourse's water is not available to be diverted into a canal, a reservoir may be built to guarantee the water level in the canal; for example, where a canal climbs to cross a range of hills through locks.

Recreation

Reservoirs often provide for recreational uses. Most reservoirs are built for a civic purpose, but still allow fishing, boating, and other activities. At most reservoirs, special rules apply for the safety of the public.

MODELLING RESERVOIR MANAGEMENT

There is a wide variety of software for modelling reservoirs, from the specialist Dam Safety Programme Management Tools (DSPMT) to the relatively simple WAFLEX, to integrated models like the Water Evaluation And Planning system (WEAP) that place reservoir operations in the context of system-wide demands and supplies.

Dry climate and water scarcity in India led to early development of water management techniques, including the building of a reservoir at Girnar in 3000 BC. Artificial lakes dating to the 5th century BC have been found in ancient Greece. An artificial lake in present-day Madhya Pradesh province of India, constructed in the 11th century, covered 650 square meters.

Drought

A drought is an extended period of months or years when a region notes a deficiency in its water supply. Generally, this occurs when a region receives consistently below average precipitation. It can have a substantial impact on the ecosystem and agriculture of the affected region. Although droughts can persist for several years, even a short, intense drought can cause significant damage and harm the local economy.

This global phenomenon has a widespread impact on agriculture. The United Nations estimates that an area of fertile soil the size of Ukraine is lost every year because of drought, deforestation, and climate instability. Lengthy periods of drought have long been a key trigger for mass migration and played a key role in a number of ongoing migrations and other humanitarian crises in the Horn of Africa and the Sahel.

Implications

Drought is a normal, recurring feature of the climate in most parts of the world. It is among the earliest documented climatic events, present in the Epic of Gilgamesh and tied to the biblical story of Joseph's arrival in and the later Exodus from Ancient Egypt. Hunter-gatherer migrations in 9,500BC Chile have been linked to the phenomenon, as has the exodus of early man out of Africa and into the rest of the world around 135,000 years ago. Modern peoples can effectively mitigate much of the impact of drought through irrigation and crop rotation.

Failure to develop adequate drought mitigation strategies carries a grave human cost in the modern era, exacerbated by ever-increasing population densities. Recurring droughts leading to desertification in the Horn of Africa have created grave ecological catastrophes, prompting massive food shortages, still recurring.

To the north-west of the Horn, the Darfur conflict in neighboring Sudan, also affecting Chad, was fueled by decades of drought; combination of drought, desertification and overpopulation are among the causes of the Darfur conflict, because the Arab Baggara nomads searching for water have to take their livestock further south, to land mainly occupied by non-Arab farming peoples.

According to a UN climate report, the Himalayan glaciers that are the sources of Asia's biggest rivers - Ganges, Indus, Brahmaputra, Yangtze, Mekong, Salween and Yellow - could disappear by 2035 due to global warming. Approximately 2.4 billion people live in the drainage basin of the Himalayan rivers.

India, China, Pakistan, Bangladesh, Nepal and Myanmar could experience floods followed by droughts in coming decades. Drought in India affecting the Ganges is of particular concern, as it provides drinking water and agricultural irrigation for more than 500 million people. The west coast of North America, which gets much of its water from glaciers in mountain ranges such as the Rocky Mountains and Sierra Nevada, also would be affected.

In 2005, parts of the Amazon basin experienced the worst drought in 100 years. A 23 July 2006 article reported Woods Hole Research Centre results showing that the forest in its present form could survive only three years of drought. Scientists at the Brazilian National Institute of Amazonian Research argue in the article that this drought response, coupled with the effects of deforestation on regional climate, are pushing the rainforest towards a "tipping point" where it would irreversibly start to die.

It concludes that the rainforest is on the brink of being turned into savanna or desert, with catastrophic consequences for the world's climate. According to the WWF, the combination of climate change and deforestation increases the drying effect of dead trees that fuels forest fires.

By far the largest part of Australia is desert or semi-arid lands commonly known as the outback. A 2005 study by Australian and American researchers investigated the desertification of the interior, and suggested that one explanation was related to human settlers who arrived about 50,000 years ago. Regular burning by these settlers could have prevented monsoons from reaching interior Australia.

In June 2008 it became known that an expert panel had warned of long term, maybe irreversible, severe ecological damage for the whole Murray-Darling basin if it does not receive sufficient water by October.

Australia could experience more severe droughts and they could become more frequent in the future, a government-commissioned report said on July 6, 2008. The Australian of the year 2007, environmentalist Tim Flannery, predicted that unless it made drastic changes, Perth in Western Australia could become the world's first ghost metropolis, an abandoned city with no more water to sustain its population.

Causes

Generally, rainfall is related to the amount of water vapour in the atmosphere, combined with the upward forcing of the air mass containing that water vapour. If either of these are reduced,the result is a drought. This can be triggered by an above average prevalence of high pressure systems, winds carrying continental, rather than oceanic air masses (ie. reduced water content), and ridges of high pressure areas form with behaviors which prevent or restrict the developing of thunderstorm activity or rainfall over one certain region.

Oceanic and atmospheric weather cycles such as the El Niño-Southern Oscillation (ENSO) make drought a regular recurring feature of the Americas along the Pacific coast and Australia. Guns, Germs, and Steel author Jared Diamond sees the stark impact of the multi-year ENSO cycles on Australian weather patterns as a key reason that Australian aborigines remained a hunter-gatherer society rather than adopting agriculture.

Human activity can directly trigger exacerbating factors such as overfarming, excessive irrigation, Deforestation, and erosion adversely impact the ability of the land to capture and hold water. While these tend to be relatively isolated in their scope, activities resulting in global climate change are expected to trigger droughts with a substantial impact on agriculture throughout the world, and especially in developing nations.

Overall, global warming will result in increased world rainfall. Along with drought in some areas, flooding and erosion will increase in others. Paradoxically, some proposed solutions to global warming that focus on more active techniques, solar radiation management through the use of a space sunshade for one, may also carry with them increased chances of drought.

Consequences

Periods of drought can have significant environmental, agricultural, health, economic and social consequences. The effect varies according to vulnerability. For example, subsistence farmers are more likely to migrate during drought because they do not have alternative food sources. Areas with populations that depend on subsistence farming as a major food source are more vulnerable to drought-triggered famine.

Drought is rarely if ever the sole cause of famine; socio-political factors such as extreme widespread poverty play a major role. Drought can also reduce water quality, because lower water flows reduce dilution of pollutants and increase contamination of remaining water sources. A few common consequences of drought include:

- Diminished crop growth or yield productions and carrying capacity for livestock.
- Dust bowls, themselves a sign of erosion, which further erode the landscape.
- Dust storms, when drought hits an area suffering from desertification and erosion.
- Famine due to lack of water for irrigation.
- Habitat damage, affecting both terrestrial and aquatic wildlife.
- Malnutrition, dehydration and related diseases.
- Mass migration, resulting in internal displacement and international refugees.
- Reduced electricity production due to insufficient available coolant for power stations; and reduced water flow through hydroelectric dams.
- Shortages of water for industrial users.
- Snakes migration and increases in snakebites.
- Social unrest.

- War over natural resources, including water and food.
- Wildfires, such as Australian bushfires, are more common during times of drought.

TYPES OF DROUGHT

As a drought persists, the conditions surrounding it gradually worsen and its impact on the local population gradually increases. People tend to define droughts in three main ways:

- Meteorological drought is brought about when there is a prolonged period with less than average precipitation. Meteorological drought usually precedes the other kinds of drought.
- Agricultural droughts are droughts that affect crop production or the ecology of the range. This condition can also arise independently from any change in precipitation levels when soil conditions and erosion triggered by poorly planned agricultural endeavors cause a shortfall in water available to the crops. However, in a traditional drought, it is caused by an extended period of below average precipitation.
- Hydrological drought is brought about when the water reserves available in sources such as aquifers, lakes and reservoirs fall below the statistical average. Hydrological drought tends to show up more slowly because it involves stored water that is used but not replenished. Like an agricultural drought, this can be triggered by more than just a loss of rainfall. For instance, Kazakhstan was recently awarded a large amount of money by the World Bank to restore water that had been diverted to other nations from the Aral Sea under Soviet rule. Similar circumstances also place their largest lake, Balkhash, at risk of completely drying out.

Mitigation Strategies

- Cloud seeding - an artificial technique to induce rainfall.
- Desalination of sea water for irrigation or consumption.
- Drought monitoring - Continuous observation of rainfall levels and comparisons with current usage levels can help prevent man-made drought. For instance, analysis of water usage in Yemen has revealed that their water table (underground water level) is put at grave risk by over-use to fertilize their Khat crop. Careful monitoring of moisture levels can also help predict increased risk for wildfires, using such

metrics as the Keetch-Byram Drought Index or Palmer Drought Index.

- Land use - Carefully planned crop rotation can help to minimize erosion and allow farmers to plant less water-dependent crops in drier years.
- Rainwater harvesting - Collection and storage of rainwater from roofs or other suitable catchments.
- Recycled water - Former wastewater (sewage) that has been treated and purified for reuse.
- Transvasement - Building canals or redirecting rivers as massive attempts at irrigation in drought-prone areas.
- Outdoor water-use restriction - Regulating the use of sprinklers, hoses or buckets on outdoor plants, the washing of motor vehicles or other outdoor hard surfaces (including roofs and paths), topping up of swimming pools.

HYDROELECTRICITY

Hydroelectricity is electricity generated by hydropower, i.e., the production of power through use of the gravitational force of falling or flowing water. It is the most widely used form of renewable energy. Once a hydroelectric complex is constructed, the project produces no direct waste, and has a considerably lower output level of the greenhouse gas carbon dioxide (CO_2) than fossil fuel powered energy plants. Worldwide, hydroelectricity supplied an estimated 816 GWe in 2005. This was approximately 20% of the world's electricity, and accounted for about 88% of electricity from renewable sources.

Electricity Generation

Most hydroelectric power comes from the potential energy of dammed water driving a water turbine and generator. In this case the energy extracted from the water depends on the volume and on the difference in height between the source and the water's outflow. This height difference is called the head. The amount of potential energy in water is proportional to the head. To obtain very high head, water for a hydraulic turbine may be run through a large pipe called a penstock.

Pumped storage hydroelectricity produces electricity to supply high peak demands by moving water between reservoirs at different elevations. At times of low electrical demand, excess generation capacity is used to pump water into the higher reservoir. When there is higher demand, water is released back into the lower reservoir through a turbine. Pumped storage schemes currently provide the only commercially important means of large-scale grid energy storage and improve the daily load factor of the generation system.

Hydroelectric plants with no reservoir capacity are called run-of-the-river plants, since it is not then possible to store water. A tidal power plant makes use of the daily rise and fall of water due to tides; such sources are highly predictable, and if conditions permit construction of reservoirs, can also be dispatchable to generate power during high demand periods.

Less common types of hydro schemes use water's kinetic energy or undammed sources such as undershot waterwheels.

A simple formula for approximating electric power production at a hydroelectric plant is: $P = hrgk$, where P is Power in kilowatts, h is height in meters, r is flow rate in cubic meters per second, g is acceleration due to gravity of 9.8 m/s, and k is a coefficient of efficiency ranging from 0 to 1. Efficiency is often higher with larger and more modern turbines.

Annual electric energy production depends on the available water supply. In some installations the water flow rate can vary by a factor of 10:1 over the course of a year.

Industrial Hydroelectric Plants

While many hydroelectric projects supply public electricity networks, some are created to serve specific industrial enterprises. Dedicated hydroelectric projects are often built to provide the substantial amounts of electricity needed for aluminium electrolytic plants, for example. In the Scottish Highlands there are examples at Kinlochleven and Lochaber, constructed during the early years of the 20th century.

The Grand Coulee Dam, long the world's largest, switched to support Alcoa aluminum in Bellingham, Washington for America's World War II airplanes before it was allowed to provide irrigation and power to citizens (in addition to aluminum power) after the war. In Suriname, the Brokopondo Reservoir was constructed to provide electricity for the Alcoa aluminium industry. New Zealand's Manapouri Power Station was constructed to supply electricity to the aluminium smelter at Tiwai Point. As of 2007 the Kárahnjúkar Hydropower Project in Iceland remains controversial.

Small-Scale Hydro-Electric Plants

Although large hydroelectric installations generate most of the world's hydroelectricity, some situations require small hydro plants.

These are defined as plants producing up to 10 megawatts, or projects up to 30 megawatts in North America. A small hydro plant may be connected to a distribution grid or may provide power only to an isolated community or a single home. Small hydro projects generally do not require the protracted economic, engineering and environmental studies associated with large projects, and often can be completed much more quickly.

A small hydro development may be installed along with a project for flood control, irrigation or other purposes, providing extra revenue for project costs. In areas that formerly used waterwheels for milling and other purposes, often the site can be redeveloped for electric power production, possibly eliminating the new environmental impact of any demolition operation. Small hydro can be further divided into mini-hydro, units around 1 MW in size, and micro hydro with units as large as 100 kW down to a couple of kW rating.

Small hydro schemes are particularly popular in China, which has over 50% of world small hydro capacity.

Small hydro units in the range 1 MW to about 30 MW are often available from multiple manufacturers using standardized "water to wire" packages; a single contractor can provide all the major mechanical and electrical equipment (turbine, generator, controls, switchgear), selecting from several standard designs to fit the site conditions. Micro hydro projects use a diverse range of equipment; in the smaller sizes industrial centrifugal pumps can be used as turbines, with comparatively low purchase cost compared to purpose-built turbines.

ADVANTAGES

Economics

The major advantage of hydroelectricity is elimination of the cost of fuel. The cost of operating a hydroelectric plant is nearly immune to increases in the cost of fossil fuels such as oil, natural gas or coal, and no imports are needed.

Hydroelectric plants also tend to have longer economic lives than fuel-fired generation, with some plants now in service which were built 50 to 100 years ago. Operating labour cost is also usually low, as plants are automated and have few personnel on site during normal operation.

Where a dam serves multiple purposes, a hydroelectric plant may be added with relatively low construction cost, providing a useful revenue stream to offset the costs of dam operation. It has been calculated that the sale of electricity from the Three Gorges Dam will cover the construction costs after 5 to 8 years of full generation.

Greenhouse Gas Emissions

Since hydroelectric dams do not burn fossil fuels, they do not directly produce carbon dioxide (a greenhouse gas). While some carbon dioxide is produced during manufacture and construction of the project, this is a tiny fraction of the operating emissions of equivalent fossil-fuel electricity generation.

Related Activities

Reservoirs created by hydroelectric schemes often provide facilities for water sports, and become tourist attractions in themselves. In some countries, aquaculture in reservoirs is common. Multi-use dams installed for irrigation support agriculture with a relatively constant water supply. Large hydro dams can control floods, which would otherwise affect people living downstream of the project.

DISADVANTAGES

Environmental Damage

Hydroelectric projects can be disruptive to surrounding aquatic ecosystems both upstream and downstream of the plant site. For instance, studies have shown that dams along the Atlantic and Pacific coasts of North America have reduced salmon populations by preventing access to spawning grounds upstream, even though most dams in salmon habitat have fish ladders installed. Salmon spawn are also harmed on their migration to sea when they must pass through turbines.

This has led to some areas transporting smolt downstream by barge during parts of the year. In some cases dams have been demolished (for example the Marmot Dam demolished in 2007) because of impact on fish. Turbine and power-plant designs that are easier on aquatic life are an active area of research. Mitigation measures such as fish ladders may be required at new projects or as a condition of re-licensing of existing projects.

Generation of hydroelectric power changes the downstream river environment. Water exiting a turbine usually contains very little suspended sediment, which can lead to scouring of river beds and loss of riverbanks. Since turbine gates are often opened intermittently, rapid or even daily fluctuations in river flow are observed. For example, in the Grand Canyon, the daily cyclic flow variation caused by Glen

Canyon Dam was found to be contributing to erosion of sand bars. Dissolved oxygen content of the water may change from pre-construction conditions.

Depending on the location, water exiting from turbines is typically much warmer than the pre-dam water, which can change aquatic faunal populations, including endangered species, and prevent natural freezing processes from occurring. Some hydroelectric projects also use canals to divert a river at a shallower gradient to increase the head of the scheme. In some cases, the entire river may be diverted leaving a dry riverbed. Examples include the Tekapo and Pukaki Rivers.

Greenhouse Gas Emissions

The reservoirs of power plants in tropical regions may produce substantial amounts of methane and carbon dioxide. This is due to plant material in flooded areas decaying in an anaerobic environment, and forming methane, a very potent greenhouse gas.

According to the World Commission on Dams report, where the reservoir is large compared to the generating capacity (less than 100 watts per square metre of surface area) and no clearing of the forests in the area was undertaken prior to impoundment of the reservoir, greenhouse gas emissions from the reservoir may be higher than those of a conventional oil-fired thermal generation plant.

Although these emissions represent carbon already in the biosphere, not fossil deposits that had been sequestered from the carbon cycle, there is a greater amount of methane due to anaerobic decay, causing greater damage than would otherwise have occurred had the forest decayed naturally.

In boreal reservoirs of Canada and Northern Europe, however, greenhouse gas emissions are typically only 2% to 8% of any kind of conventional fossil-fuel thermal generation. A new class of underwater logging operation that targets drowned forests can mitigate the effect of forest decay.

In 2007, International Rivers accused hydropower firms for cheating with fake carbon credits under the Clean Development Mechanism (CDM), for hydropower projects already finished or under construction at the moment they applied to join the CDM. These carbon credits – of hydropower projects under the CDM in developing countries – can be sold to companies and governments in rich countries, in order to comply with the Kyoto protocol.

Population Relocation

Another disadvantage of hydroelectric dams is the need to relocate the people living where the reservoirs are planned. In February 2008, it was estimated that 40-80 million people worldwide had been physically displaced as a direct result of dam construction.

In many cases, no amount of compensation can replace ancestral and cultural attachments to places that have spiritual value to the displaced population. Additionally, historically and culturally important sites can be flooded and lost. Such problems have arisen at the Three Gorges Dam project in China, the Clyde Dam in New Zealand and the Ilýsu Dam in Southeastern Turkey.

Dam Failures

Failures of large dams, while rare, are potentially serious — the Banqiao Dam failure in Southern China resulted in the deaths of 171,000 people and left millions homeless. Dams may be subject to enemy bombardment during wartime, sabotage and terrorism. Smaller dams and micro hydro facilities are less vulnerable to these threats. The creation of a dam in a geologically inappropriate location may cause disasters like the one of the Vajont Dam in Italy, where almost 2000 people died, in 1963.

Affected by Flow Shortage

Changes in the amount of river flow will correlate with the amount of energy produced by a dam. Because of global warming, the volume of glaciers has decreased, such as the North Cascades glaciers, which have lost a third of their volume since 1950, resulting in stream flows that have decreased by as much as 34%. The result of diminished river flow can be power shortages in areas that depend heavily on hydroelectric power.

Comparison with other Methods of Power Generation

Hydroelectricity eliminates the flue gas emissions from fossil fuel combustion, including pollutants such as sulfur dioxide, nitric oxide, carbon monoxide, dust, and mercury in the coal. Hydroelectricity also avoids the hazards of coal mining and the indirect health effects of coal emissions. Compared to nuclear power, hydroelectricity generates no nuclear waste, has none of the dangers associated with uranium mining, nor nuclear leaks. Unlike uranium, hydroelectricity is also a renewable energy source.

Compared to wind farms, hydroelectricity power plants have a more predictable load factor. If the project has a storage reservoir, it can be dispatched to generate power when needed. Hydroelectric plants can be easily regulated to follow variations in power demand.

Unlike fossil-fueled combustion turbines, construction of a hydroelectric plant requires a long lead-time for site studies, hydrological studies, and environmental impact assessment. Hydrological data up to 50 years or more is usually required to determine the best sites and operating regimes for a large hydroelectric plant.

Unlike plants operated by fuel, such as fossil or nuclear energy, the number of sites that can be economically developed for hydroelectric production is limited; in many areas the most cost effective sites have already been exploited. New hydro sites tend to be far from population centers and require extensive transmission lines. Hydroelectric generation depends on rainfall in the watershed, and may be significantly reduced in years of low rainfall or snowmelt. Long-term energy yield may be affected by climate change.

Utilities that primarily use hydroelectric power may spend additional capital to build extra capacity to ensure sufficient power is available in low water years. In parts of Canada (the provinces of British Columbia, Manitoba, Ontario, Quebec, Newfoundland and Labrador) hydroelectricity is used so extensively that the word "hydro" is often used to refer to any electricity delivered by a power utility.

The government-run power utilities in these provinces are called BC Hydro, Manitoba Hydro, Hydro One (formerly "Ontario Hydro"), Hydro-Québec and Newfoundland and Labrador Hydro respectively. Hydro-Québec is the world's largest hydroelectric generating company, with a total installed capacity (2007) of 35,647 MW, including 33,305 MW of hydroelectric generation.

Countries with the most Hydro-Electric Capacity

The ranking of hydro-electric capacity is either by actual annual energy production or by installed capacity power rating. A hydro-electric plant rarely operates at its full power rating over a full year; the ratio between annual average power and installed capacity rating is the capacity factor.

The installed capacity is the sum of all generator nameplate power ratings. Sources came from *BP Statistical Review - Full Report 2007* List of the largest hydoelectric power stations. Norway produces 98-99% of its electricity from hydroelectric.

Brazil, Canada, Norway and Venezuela are the only countries in the world where the majority of their internal electric energy production is from hydroelectric power.

Index

F

G

H

I

L

M

N

O

P